T0137974

Studies in Systems, Decision and Control

Volume 176

Series editor

Janusz Kacprzyk, Polish Academy of Sciences, Warsaw, Poland
e-mail: kacprzyk@ibspan.waw.pl

The series "Studies in Systems, Decision and Control" (SSDC) covers both new developments and advances, as well as the state of the art, in the various areas of broadly perceived systems, decision making and control–quickly, up to date and with a high quality. The intent is to cover the theory, applications, and perspectives on the state of the art and future developments relevant to systems, decision making, control, complex processes and related areas, as embedded in the fields of engineering, computer science, physics, economics, social and life sciences, as well as the paradigms and methodologies behind them. The series contains monographs, textbooks, lecture notes and edited volumes in systems, decision making and control spanning the areas of Cyber-Physical Systems, Autonomous Systems, Sensor Networks, Control Systems, Energy Systems, Automotive Systems, Biological Systems, Vehicular Networking and Connected Vehicles, Aerospace Systems, Automation, Manufacturing, Smart Grids, Nonlinear Systems, Power Systems, Robotics, Social Systems, Economic Systems and other. Of particular value to both the contributors and the readership are the short publication timeframe and the world-wide distribution and exposure which enable both a wide and rapid dissemination of research output.

More information about this series at http://www.springer.com/series/13304

Edy Portmann · Marco E. Tabacchi
Rudolf Seising · Astrid Habenstein
Editors

Designing Cognitive Cities

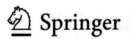 Springer

Editors
Edy Portmann
Human-IST Institute
University of Fribourg
Fribourg, Switzerland

Marco E. Tabacchi
Istituto Nazionale di Ricerche
  Demopolis
Demopolis, Italy

Rudolf Seising
Deutsches Museum München
The Research Institute for the History
  of Science and Technology
Munich, Germany

Astrid Habenstein
Transdisciplinary Research Centre Smart
  Swiss Capital Region (TRCSSCR)
University of Bern
Bern, Switzerland

ISSN 2198-4182   ISSN 2198-4190   (electronic)
Studies in Systems, Decision and Control
ISBN 978-3-030-13102-9   ISBN 978-3-030-00317-3   (eBook)
https://doi.org/10.1007/978-3-030-00317-3

This Springer imprint is published by the registered company Springer Nature Switzerland AG
The registered company address is: Gewerbestrasse 11, 6330 Cham, Switzerland

Foreword

It is not easy to write a meaningful preface to a book which is as ambitious and as unconventional as this book is. Its focus is cognitive science and soft computing—a sphere of human cognition which was born with the computer age. Around the middle of last century, I witnessed the birth. The source of excitement was Artificial Intelligence (AI). There were many exaggerated expectations which were promoted by the AI leaders John McCarthy, Marvin Minsky, Allen Newell, and others. I was just beginning my teaching career and was enthuse by what I saw and heard. My first paper was "Thinking machines—a new field in electrical engineering" published in January 1950 (Zadeh 1950), in Columbia Engineering Quarter. In this paper, I started with a list of headlines which appear in the popular press at that time, for example, "An electrical machine capable of translating foreign languages is being built." What is remarkable is that today, many years later, we still do not have machines which are close to delivering, close to human quality machine translation. Exaggerated expectations were fueled by quest by financial support. It was a game which AI paid dearly with many long month of winter from which it is beginning to emerge.

In my view, the pioneers of AI made a major mistake. They put all of the AI's eggs into the basket of classical logic and turning away from anything that was not logic base. What they overlooked was that much of everyday human reasoning—the reasoning which underlies cognitive science and soft computing is not based on classical logic. In particular, fuzzy logic on approximate reasoning plays essential roles in human reasoning and human cognition. For many years, the AI community took a very skeptical position on anything that was not based on classical logic—fuzzy logic numerical computations. Neurocomputing, evolutionary computing, and probability theory—methodologies which are at the center of what successful AI is today. For me, the best example of successful AI is the "smart" phone. The smart phone is indeed a remarkable product which was science fiction not that long ago.

Approximate reasoning underlies much of human reasoning, cognitive science, and soft computing. In approximate reasoning, the optic of reasoning and computation are for the most part is fuzzy sets, that is, classes which unsharp (fuzzy) borders. This is what traditional AI overlooked. It will take many Ph.D. theses on over a long period of time to develop what could be justifiably called a unified theory of inexact and approximate reasoning. Today, all we have are fragments. The mathematics of approximate reasoning is much more complex than the mathematics exact reasoning. In fact, the mathematics of inexact and approximate reasoning will be a new kind of mathematics. It would be much more computer-oriented than what mathematics is today.

I should like to put on the table a new concept—invaluent variable. In large measure, computation involves assignment of values to variables. In realistic settings, there are many situations in which we cannot assign a value to a variable X because we do not know what the value is and do not have a clear idea how it can be define, for example, variables which related to fairness, rationality, beauty, and relevance. Variables of this kind are invaluent variables, and the underlying issue is invaluence. Invaluence is pervasive in cognitive science, cognitive computer, and approximate and inexact reasoning. In the context of invaluency in large measure, we are dealing with perceptions rather than measurements. If X is an invaluent variable, a value which can be assign to X is a Z-number (Zadeh 2011). A Z-number is a construct with three components. The first component is the name of the variable. The second component called "value" is an estimated value of X, based mostly on perception-based information. The third component called "confidence" is a fuzzy number which is an estimate of the goodness/correctness/reliability of A as an esti-mate of X, for example, first component: projected deficit; second component: about $ 5,000,000; third component: high is a linguistic description of a fuzzy number. For simplicity, linguistic fuzzy numbers are assumed to be high, medium, and low. Fuzzy numbers can be computed with a recent groundbreaking monograph by Rafik (2016).

In coming years, I envisage that theories of fuzzy numbers and their applications will grow invisibility and importance. It may turn out that the concept of a Z-number is what is needed to develop cognitive science, cognitive computer, and inexact and approximate reasoning. The importance of Z-numbers derives from the fact that they can be computed with and reason with logically.

Berkeley, CA, USA   Lotfi Zadeh
October 2016   Professor emeritus and Director
                                         Berkeley Initiative in Soft Computing (BISC)[1]

References

Rafik A (2016) Arithmetic operations and fuzzy numbers. World Scientific, INC
Zadeh LA (1950) Thinking machines—a new field in electrical engineering. Columbia Eng Q
    12–30
Zadeh LA (2011) A note on Z-numbers. Inf Sci 181(14):2923–2932

---

[1] The author of this Foreword, Professor emeritus Dr. Lotfi A. Zadeh, world-renowned computer scientist and source of the author's and editor's curiosity, died on September 06, 2017, at the age of 96. He will be greatly missed. His obituary at UC Berkeley may be found at: https://engineering. berkeley.edu/2017/09/remembering-lotfi-zadeh. Accessed March 01, 2018.

Contents

Editors and Contributors

About the Editors

Edy Portmann is a researcher and scholar, specialist and consultant for semantic search, social media, and soft computing. Currently, he works as a Swiss Post-Funded Professor of Computer Science at the Human-IST Institute of the University of Fribourg, Switzerland. He studied for a B.Sc. in Information Systems at the Lucerne University of Applied Sciences and Arts, for an M.Sc. in Business and Economics at the University of Basel, and for a Ph.D. in Computer Sciences at the University of Fribourg. He was a Visiting Research Scholar at National University of Singapore (NUS), Postdoctoral Researcher at University of California at Berkeley, USA, and Assistant Professor at the University of Bern. Next to his studies, he worked several years in a number of organizations in study-related disciplines. Among others, he worked as Supervisor at Link Market Research Institute, as Contract Manager for Swisscom Mobile, as Business Analyst for PwC, as IT Auditor at Ernst & Young and, in addition to his doctoral studies, as Researcher at the Lucerne University of Applied Science and Arts. Edy Portmann is repeated nominee for Marquis Who's Who, selected member of the Heidelberg Laureate Forum, co-founder of Mediamatics, and co-editor of the Springer Series "Fuzzy Management Methods," as well as author of several popular books in his field. He lives happily married in Bern and has three lively kids.

Marco E. Tabacchi is currently the Scientific Director at Istituto Nazionale di Ricerche Demopolis, Italy. He is a transdisciplinary scientist, with interests in cognitive science, computational intelligence applied to IoT, and philosophy and history of soft computing.

Rudolf Seising obtained his Ph.D. in Philosophy of Science and his Habilitation in History of Science from the Ludwig Maximilian University of Munich, Germany, after studies of Mathematics, Physics, and Philosophy at the Ruhr-University Bochum, Germany. He has been Scientific Assistant for Computer Sciences (1988–1995) and for History of Sciences (1995–2002) at the University of the Armed

Forces in Munich. From 2002 to 2008, he was with the Core unit for Medical Statistics and Informatics at the University of Vienna Medical School, Austria. He acted as Professor for History of Science at the Friedrich-Schiller-University Jena, Germany, and at the Ludwig Maximilian University of Munich, Germany. He works now at the Research Institute for the History of Science and Technology at the Deutsches Museum in Munich and is Lecturer at the Faculty of History and Arts at the Ludwig Maximilian University of Munich. He was Visiting Researcher (2008–2010) and Adjoint Researcher (2010–2014) at the European Centre for Soft Computing in Mieres (Asturias), Spain, and he has been several times Visiting Scholar at the University of California in Berkeley, USA. He is Chairman of the IFSA Special Interest Group "History" and of the EUSFLAT Working Group "Philosophical Foundations." He is member of the IEEE Computational Intelligence Society (CIS) History Committee and of the IEEE CIS Fuzzy Technical Committee. His main areas of research comprise the historical and philosophical foundations of science and technology.

Astrid Habenstein leads the Transdisciplinary Research Centre Smart Swiss Capital Region (TRCSSCR) at the University of Bern, Switzerland. She graduated at the University of Bielefeld, Germany, in History and Philosophy (2005) and received her Ph.D. in Ancient History at the University of Bern (2012). Astrid Habenstein worked as Research Assistant at the University of Bern and as Lecturer at the Universities of Bern and Basel. She was member of Edy Portmann's project on Smart and Cognitive Cities at the Institute of Information Systems, University of Bern (2016–2017). Her main interdisciplinary and transdisciplinary research interests include Smart and Cognitive Cities, sociological theory formation (knowledge research, theories of design, systems theory, theories of interaction), and research development.

Contributors

José M. Alonso Centro Singular de Investigación en Tecnoloxías da Información (CiTIUS), Universidade de Santiago de Compostela, Rúa de Jenaro de la Fuente Domínguez, Santiago de Compostela, Spain

Kurt Bollacker Stitch Fix Inc, San Francisco, CA, USA

Ciro Castiello Department of Informatics, University of Bari "Aldo Moro", Bari, Italy

Javier Cuenca Faculty of Engineering, Mondragon University, Arrasate-Mondragon, Spain

Edward Curry Insight Centre for Data Analytics, National University of Ireland Galway, Galway, Ireland

Natalia Díaz-Rodríguez Stitch Fix Inc, San Francisco, CA, USA

Sara D'Onofrio Human-IST Institute, University of Fribourg, Fribourg, Switzerland

Luka Eciolaza Faculty of Engineering, Mondragon University, Arrasate-Mondragon, Spain

Astrid Habenstein Transdisciplinary Research Centre Smart Swiss Capital Region (TRCSSCR), University of Bern, Bern, Switzerland

Miroslav Hudec Faculty of Organizational Sciences, University of Belgrade, Belgrade, Serbia

Patrick Kaltenrieder University of Bern, Bern, Switzerland

Nasrin Khansari Electrical & Systems Engineering Department, University of Pennsylvania, Philadelphia, PA, USA

Thomas Krebs University of Bern, Bern, Switzerland

Felix Larrinaga Faculty of Engineering, Mondragon University, Arrasate-Mondragon, Spain

Xian Li Stitch Fix Inc, San Francisco, CA, USA

Mo Mansouri School of Systems and Enterprises, Stevens Institute of Technology, Hoboken, NJ, USA

Corrado Mencar Department of Informatics, University of Bari "Aldo Moro", Bari, Italy

Thomas Myrach University of Bern, Bern, Switzerland

Elpiniki Papageorgiou Department of Electrical Engineering, University of Applied Sciences (TEI) of Thessaly, Larisa, Greece

Jorge Parra University of Bern, Bern, Switzerland

Edy Portmann Human-IST Institute, University of Fribourg, Fribourg, Switzerland

Rudolf Seising The Research Institute for the History of Science and Technology, Deutsches Museum München, Munich, Germany

Alejandro Sobrino University of Santiago de Compostela, Santiago de Compostela, Spain

Marco E. Tabacchi Istituto Nazionale Di Ricerche Demopolis, Demopolis, Italy

Enric Trillas University of Oviedo, Oviedo, Spain

Noémie Zurlinden University of Bern, Bern, Switzerland

Part I
Introduction

Designing Cognitive Cities

Marco E. Tabacchi, Edy Portmann, Rudolf Seising
and Astrid Habenstein

Abstract The following text intends to give an introduction into some of the basic ideas which determined the conception of this book. Thus, the first part of this article introduces the terms "City", "Smart City" and "Cognitive City". The second part gives an overview of design theories and approaches such as Action Design Research and Ontological Design (a concept in-the-making), in order to deduce from a theoretical point of view some of the principles that needs to be taken into account when designing the Cognitive City. The third part highlights some concrete techniques that can be usefully applied to the problem of citizen communication for Cognitive Cities (namely Metaheuristics, Fuzzy Sets and Fuzzy Logic, Computing with Words, Computational Intelligence Classifiers, and Fuzzy-based Ontologies). Finally, we introduce the articles of this book.

Keywords Action Design Research (ADR) · Cognitive city
Collective intelligence—urban intelligence · Computational intelligence
Citizen communication · Computing with words · Fuzzy logic
Ontological design · Smart city

M. E. Tabacchi
Istituto Nazionale Di Ricerche Demopolis, Demopolis, Italy
e-mail: marcoelio.tabacchi@unipa.it

E. Portmann
Human-IST Institute, University of Fribourg, Fribourg, Switzerland
e-mail: edy.portman@unifr.ch

R. Seising
The Research Institute for the History of Science and Technology, Deutsches Museum München, Munich, Germany
e-mail: r.seising@deutsches-museum.de

A. Habenstein (✉)
Transdisciplinary Research Centre Smart Swiss Capital Region (TRCSSCR), University of Bern, Bern, Switzerland
e-mail: astrid.habenstein@histdek.unibe.ch

© Springer Nature Switzerland AG 2019   3
E. Portmann et al. (eds.), *Designing Cognitive Cities*, Studies in Systems,
Decision and Control 176, https://doi.org/10.1007/978-3-030-00317-3_1

1 From Smart to Cognitive Cities

1.1 The City in the Age of Globalization

The city is (and always was) a complex, multifaceted phenomenon. There are numerous scientific approaches and disciplines dealing with its dimensions and, accordingly, widely varying ideas about what constitutes the city (Mieg 2013). Not least due to this fact, it seems almost impossible to give the term "city" one universal, scientifically substantiated and comprehensive definition—and perhaps it does not even make sense to try to find one as this would be rather superficial (Eckardt 2014). However, most definitions and description have in common that they usually associate or postulate two general perspectives as central to describe the city appropriately, thus describing the reference framework and fields of action for a sustainable urban development: the spatial-material and the social-cultural dimension. In a very general sense, cities are almost always considered as topographically describable and geographically definable places with characteristic, condensed settlements and infrastructures, which separate the city from the non-urban surrounding area (with fluent transitions); they are places where a large number of different people live, work, create specific forms of life and, despite all heterogeneity, also develop a common identity (Mieg 2013).

In this sense, the city has been for millennia a settlement structure and way of life which is typical and of great importance to mankind. And it gets even more important: Since the beginning of the 20th century, the world population has quadrupled to about 7.4 billion people in 2015, and in its most recent population report, the UN estimates that in 2030 ca. 8.5 billion and in 2050 ca. 9.1 are going to live on planet Earth (United Nations 2015). This development is accompanied by a process of accelerated urbanization: In 1950 more than 70% of the world's population still lived in rural regions. In 2007, for the first time the urban population was larger than the rural population, and the UN's latest estimates assume that in 2050 66% of all people are going to be inhabitants of cities. The regional differences here are sometimes immense: in North and Latin America already 84%, in Europe 73%, while in Africa and Asia still less than half of the population live in cities (United Nations 2014). But the general trend is tangible everywhere.

Sometimes this is regarded as sign of progress, as life in the city can allow more spiritual, cultural and social growth, as well as better living conditions and levels of care. But as we all know this does not necessarily hold true everywhere: Today's cities struggle with the repercussions of the rapid growth of population (and simultaneously decreasing availability of living spaces) as well as globalization, climate change, the increasing scarcity of natural resources and the so-called Third (or Fourth) Industrial Revolution. So-called megacities (with more than 10 million inhabitants, such as New York, Tokyo, Manila, Beijing, Rio de Janeiro or Mumbai), whose number has risen from 10 to 28 since 1990 (United Nations 2014), are symbols of the consequences of an urbanization that is barely controllable under the conditions of globalization. And these trends are also tangible in the relatively small cities of Western Europe, though not nearly as bad as in the megacities of Asia and Latin America. European cities become larger and more complex, too, and their heavily used infrastructures are increasingly coming up against their load limits. Environmental issues; the integration of different cultures and traditions; healthcare and quality of life, especially for the elders and disabled; pollution; eco-compatible transport and working policies—these are but some of the problems modern cities have to deal with worldwide.

These developments are going to change the relationship between citizen and city immensely. Therefore, it is more important than ever before to think about how cities have to be shaped to provide their inhabitants with the means and resources for a good life. One approach to deal with the challenges of modern cities is associated with the buzzword "Smart City".

1.2 Smart City–Cognitive City

There are a lot of perspectives, research approaches and, subsequently, lots of definitions and conceptions of the term "Smart City" (Albino et al. 2015). But all of them have in common the basic idea that the enrichment of city-relevant functions with ICT can contribute to develop efficiently and sustainably the socioecological design of the urban space (Portmann and Finger 2015). The collection and analysis of city-related data as well as the coordination of their use by means of internet and web-based services are intended to help to develop cities into better, more beautiful, more viable places.

The challenges cities have to deal with and for which smart solutions proved to be especially suitable are often very similar, albeit with different focuses, depending on the specific character, problems and needs of the city in question. In general, smart solutions are applied to subjects such as smart mobility, smart energy, smart environment, smart economy, smart living, and smart governance. Hereby, Smart City-concepts and -projects tend to focus on the enhancement of efficiency and sustainability. Especially with respect to transport and mobility, (public) security, environmental and climate protection (waste management, resource-efficient use of energy and water), and municipal administration services there are impressive possibilities to make use of ICT to meet the challenges in the city (Townsend 2013).

Nonetheless, there are also critics with well-founded demurs against the Smart City. Recently, the discussion turns towards two aspects: Firstly, critics point out the problem that Smart City-initiatives are too often planned "top-down" (Cohen 2015; Dyer et al. 2017). Secondly, it becomes apparent that the focus on efficiency and sustainability is expedient and very important in many respects, but in certain cases not the best way to deal with the needs of the individual (Finger and Portmann 2016). Therefore, the legitimate claim of the citizens to participate in shaping their cities and communities and the human need to be perceived as individuals attracts more and more awareness (Dyer et al. 2017; Beinrott 2015). The Cognitive City-approach is supposed to address these requirements.

Fig. 1 Connectivism
(following Siemens 2006)

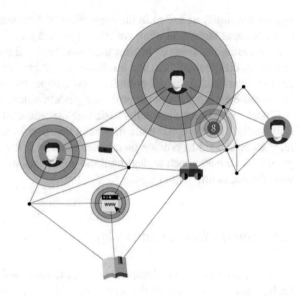

The main-characteristic of smart cities is the collection, analysis and preparation of data to obtain information that can be used to address specific problems or needs in the city (Finger and Portmann 2015). A city can become smarter by collecting high-quality data made available to the various stakeholders of a city (Hurwitz et al. 2015). Thus, they are able to better deal with specific problems or needs in the city. Therefore, it is imperative to build up and to make use of the "urban intelligence", the collective intelligence of the city.

Collective intelligences consist of individual intelligences: more or less intelligent human beings as well as more or less intelligent objects such as Artificial Intelligences (AI) or electronic devices. They form a network that is more than the sum of its elements and contributes to problem solving differently than individual intelligences would (Malone and Bernstein 2015). The "glue" that connects them can be described best as "connectivism", in reference to the eponymous learning and cognition theory by Siemens (2006). In contrast to conventional theories such as behaviorism, cognitivism and constructivism, the "Connected Learning Theories" (Caine and Caine 2011; Ito et al. 2013) and the "Incidental Connectivism" (Siemens 2006) understand learning as a process in which the learning subject—or object!—forms networks by linking to nodes. Nodes can be other people, but also databases, apps, the Internet, smartphones, books, images, etc., which have their own networks that the learning subject/object also accesses by connecting to the corresponding node (Fig. 1). Their diversity and the diversity of their networks help to generate knowledge that extends the original knowledge, or even goes beyond it (Siemens 2006). The linking of nodes occurs by interaction and communication.

Cognitive City-approaches focus on the last aspect. The term refers to a continually interconnecting and exchanging web of information and communication hubs that lies at the core of tomorrow's (and today's) cities. In the Cognitive City the human factor is added to the communication loop, and communication between persons and persons, persons and machines and machines and machines constantly happens through any available means. The technical basis are cognitive computer systems, which are capable of recognizing patterns in the huge amounts of data and learn by interacting and communicating with the people who use them (Hurwitz et al. 2015; Wilke and Portmann 2016). At the same time, they are able to learn more about what we feel, want, and need, from the constant interaction with the people who use it. In this way, new data are collected and processed. Developments such as cloud-based social feedback, crowdsourcing, and predictive analytics allow cities to actively and independently learn, build, search, and expand when new information is added to the already existing ones.

Big Data and the Internet of Things (IoT), a network of objects that are equipped with sensors, software and network connectivity, are going to play an increasingly important role here (Townsend 2013). The objects are capable to collect and forward large amounts of data cost- and energy-efficiently, as well as exchanging them autonomously among themselves. Furthermore, the objects are capable of cooperating with existing Internet infrastructures. The entire urban environment is equipped with sensors that collect data which are made available in a cloud. Thus, every public implement is at the same time a useful fixture per se, and a fast, cheap and ubiquitous mean of data gathering, based on sensors and integrating actuators and distributed intelligent components via an extended mesh of mobile and static sharing points. This creates a permanent interaction between urban residents and the surrounding technology.

Cognitive City-concepts are not supposed to and cannot replace Smart City-approaches but complement them by focusing on a specific aspect of the Smart City: interaction and communication between the stakeholders and the city. Thus, the Cognitive City is not merely another topic such as smart mobility or smart energy, but another *perspective* that affects the Smart City as a whole: Cognitive City principles (as well as techniques resp. technologies) are applicable to all issues of the Smart City if it concerns aspects of interaction and communication. As said before, Cognitive City aims to answer demands of the future cities that cannot be met by the means of efficiency and sustainability only, but also address resilience as well as the citizens need for participation and individualism. Therefore, designing cognitive cities means designing the reciprocity of communication resp. interaction between city-related ICT and the citizens.

2 Designing Cognitive Cities

2.1 Theories of Design

No other species shaped planet Earth as much as humans did. The ability to transform and thus to "design" our natural, material and social environment according to our needs is not always positive but seems to be an essential part of human nature.

Marx (1867) spoke of the "toolmaking animal", a conception that Bergson (1907) and Scheler (1926) transformed in the famous phrase "Homo faber", which has since found its way into philosophical anthropology (Ropohl 2010). However, the ideas on how to define the term "design" vary widely (Mareis 2014). It was not until the nineteenth century that it gained the basic semantic content it has today: the preparatory, modelling process and its result, the designed artefact (Walker 1989; Hirdina 2010). Ever since, a vivid theoretical discussion emerged that is characterized by expanding the object of designing processes far beyond the traditional meaning of design as professionalized forms of handicraft, respectively technical drafting. Currently, the concept encompasses a general understanding of technical and organizational planning, conceptualizing and problem solving, too (Mareis 2014). Already at the end of the 1960s, Herbert A. Simon declared in his renowned book "The Science of the Artificial": "Everyone designs who devises courses of action aimed at changing existing situations in preferred ones" (Simon 1969). This assertion is cited in virtually every publication on design theory and expresses that design can be everything that changes an instantaneous, unsatisfactory state for the better (Mareis 2014). In this sense, design is the practice of transforming the present and shaping the future.

One of the earliest debates in design theories about what design is (or should be), its purposes and principles as well as social, economic, scientific, philosophical or pragmatic implications started from the question what constitutes "good" design. The debate concentrated on the relationship between form and function; distancing from traditional concepts of aesthetics as measurement of the quality of design, it was argued that the design of an object is "good" (even beautiful), if it is functional. But since the 1960s, the technocratic and rationalistic tendency of this point of view was increasingly scrutinized. Critics argued that the rather lopsided orientation of design to questions of usefulness and functionality does not necessarily correspond to man's need. As a consequence, they called for human centered design. The focus shifted from the categories of usefulness and functionality to the perception, reception and usage of design artefacts as well as their role as carriers and intermediaries of meaning. Subsequently, theorists of design got interested in the actors of the design process as well as the impact of their social, cultural, economic or political environment (Mareis 2013). An important contribution to these discussions provided the so-called "semantic turn" in design theory (Krippendorff 2006). Emphasizing the significance of the users as active participants in the design process, Krippendorff pointed out that designers should not only reproduce their own concepts of design, aesthetics and usability, but also respect the conceptions, values and knowledge of all who are affected by the artefacts. Human centered design requires an understanding that integrates recursively the understanding of others into one's own.

And who designs? Simon (1969) declared that everyone could be a designer, not only scientists and engineers. Today, researchers go some steps further: To Rittel (1987) design is goal-oriented and reason-based problem-solving resp. decision-making. In this view, planning and thus designing is supposed to be an argumentative process that should not be conducted as closed scientific expert discourses, but participatory. Therefore, Rittel proposes—at least regarding "wicked problems" (very complex problems in the field of sociopolitical planning)—to replace the "expert model" with a "conspirative planning model" (Rittel 2013). In addition to the traditional view on designers and users as actors in design processes, theories of design that were influenced by the Actor-Network-Theory suggest taking into account the impact of inanimate artefacts as non-human actors, too (Mareis 2013). These concepts assume that artefacts have their own forms of material-visual communication and interaction that enables human and non-human actors to develop networks and thus to generate meaning and knowledge together.

Another important matter of subject in the theoretical debates is the relationship between design and knowledge, respectively the notion of design as epistemic practice. This aspect was discussed intensively since the 1960s, when Simon realized that artificiality was a prominent feature of modernity: "The world in which we live today is far more of a man-made or artificial than a natural world" (Simon 1969). Therefore, he called for the establishment of a new "science of the artificial" to deal adequately with artificially created objects and phenomena of information technologies. He regarded design as a scientific method of practical thought, planning, decision-making and anticipated that the production of scientific-technological knowledge necessarily and increasingly would take place in application- and design-oriented contexts, thus overcoming the borders between "science" and "practice". In this view, design produces specific forms of knowledge that are different from the conventional forms of knowledge production in the natural sciences or the humanities. Today, design research does not mean to explore design practice with scientific methods, but to generate new knowledge with the means and methods of design practice itself. The interplay of implicit, objectified and technical forms of procedures and knowledge forms an "epistemology of design" that focuses on design practices themselves (Mareis 2011).

To sum it up, debates on design theory are strongly influenced by the following key points: (1) technology-driven design principles versus the ideal of human centered design; (2) design as a sociocultural process; (3) the actors (the users as active participants in the design process; designing as argumentative practice that should not be conducted solely by experts; the impact of inanimate artefacts as non-human actors); (4) the 'epistemology of design' (designing produces specific forms of knowledge). All of these aspects are also relevant for designing the future city and should be part of concepts for smart und cognitive cities, too. Some of this is already discussed in information systems research, especially but not exclusively regarding design as epistemic practice. An important approach is Design Science Research (DSR), a set of synthetic and analytical techniques helping researchers to create new knowledge through designing respectively building ("knowledge through making") and analyzing the use and/or performance of artefacts along with reflection and abstraction. The aim is "to improve and understand the behavior of aspects of Information Systems" (Vaishnavi and Kuechler 2015). An interesting advancement and concrete proposal how to implement the principles of DSR is Action Design Research (ADR).

2.2 Action Design Research and Ontological Design

ADR is appreciated as a concept that tries to bring research and practice into the best possible exchange and combines design science with action research (Sein et al. 2011). In this view IT-artefact are "ensemble artefacts" originating from the interplay of design and the context in which design takes place (Gregor and Jones 2007). Thus, a research method is required that explicitly regards artefacts as ensembles "emerging from design, use, and ongoing refinement in an organizational context", which are shaped by the "interests, values and assumptions of a wide variety of communities of developers, investors and users" (Orlikowski and Iacono 2001, p. 131; Sein et al. 2011, p. 38). Building and evaluating ensemble IT artefact culminate in prescriptive design knowledge.

ADR-projects consist of four stages encompassing principles and specific tasks for every stage (Fig. 2). In Stage One ("Problem Formulation") the research question, methodology and research design are developed. Stage One is characterized by two principles: "Practice-Inspired Research" and "Theory ingrained artefact". The combination of these principles enables to connect better research and practice: By the means of scientific theory formation the solution for a concrete field-problem that is regarded as "knowledge-creation opportunity" can be described exactly and an appropriate prototype can be developed, providing the basis for cycles of "Building, Intervention and Evaluation" (BIE) in Stage Two. In Stage Two this initial design is further shaped by organizational use and subsequent design cycles. The elements of the iterative process are intertwined with constant evaluation. Reciprocal shaping, mutually influential roles, and evaluation are the leading principles of this stage to ensure this and to intensify the mutual learning of the ADR-partners.

Stage Three is about "Reflection and Learning". This phase represents the step from building a solution for a particular problem to apply the results of the research process to a broader class of problems, an important characteristic of ADR: ADR-teams do not intend to solve only one specific problem, or to intervene within the organizational context of the problem, they aim to generate knowledge that can

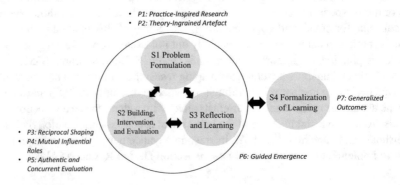

Fig. 2 Stages (S) and Principles (P) of ADR (following Sein et al. 2011, 41)

be applied to the class of problems that the specific problem exemplifies. Hereby, the ensemble artefact "will reflect not only the preliminary design created by the researchers but also its ongoing shaping by organizational use, perspectives, and participants" (Sein et al. 2011; Iivari 2007). Any Observations, changes to the artefact or anticipated as well as unanticipated consequences during the BIE iterations needs to be reflected. These observations are the basis to "Formalize the Learning" and to derive "Design Principles" in Stage Four. This phase draws on the principle that generalized outcomes are a critical component of ADR. The researchers outline the accomplishments realized in the IT-artefact and describe their results on three levels: generalizing the problem, generalizing the solution, and deriving the Design Principles that connects the generalized outcomes to a class of solutions and a class of problems.

Currently, ADR gains more and more attention and proponents, and there are first (more or less theoretical) attempts, too, to apply the idea to Smart City-projects (Maccani et al. 2014a, b; Ståhlbröst et al. 2015a, b). There are several reasons that makes ADR interesting for the Cognitive City. On the one hand, this applies to the point of view of creating a methodological and organizational framework in which different groups of vehicles are linked. A similar framework is also needed for the development of the smart and cognitive cities. There are numerous stakeholders that want to take part in the project of designing their cities. In addition, principles of transdisciplinarity could be better implemented, thus ensuring that new insights from research and development are better being disseminated and applied in the "real world". Simultaneously, the data base increases, which in turn can be used for new developments. In addition, action research elements appear particularly promising in such an application-oriented area of development and research as the Smart City. However, there remain also some difficulties, especially if ADR should be applied to Cognitive Cities.

A research approach is needed to adjust the basics of ADR to the framework that characterizes the Cognitive City. Therefore, we want to outline first theoretical considerations on principles for Ontological Design Research (ODR), a research method in the making which is based on ADR, but focusses on the particularities regarding the user, the context and the artefact in sociotechnical systems like the Cognitive City. Hereby the notion "We shape our world as this world affects and shapes us" (Willis 2006) can be considered as the basic idea of ontological design.

1. The user and the context: The primary goal of ADR is not to integrate the user into the design-process in the sense of human-centered design. ADR was originally made to improve the cooperation between research and practice by implementing design knowledge methods and focusing on the artefact and its context. This does not mean at all that the (individual) human beings are unimportant to ADR, on the contrary. But the human factor appears as one aspect among others which belong to the context of an IT-artefact. However, to retrieve the full potential of the user as co-researcher and co-designer something more needs to be done especially as the call for IT-projects that take the user into account und ask them what they want gets louder. Furthermore, it is relatively difficult to capture the so-called ADR-team in the concept. This is plausible

insofar as it is difficult to generalize the composition, roles and tasks, for this may differ very much from project to project. Unfortunately, it is even harder to capture the roles and functions of the end-user in ADR-projects. Of course, they appear in Seins BIE-schemata, but not as member of the ADR-team (Sein et al. 2011), and the description of their role and functions stays rather superficial. Next to this, it remains somewhat unclear what constitutes an "organizational context" and how to deal with the differences between the various possible contexts.

All of it is less problematic in small, well-defined organizational contexts, especially if practitioners and/or researchers are simultaneously the end-user. However, in the complex situation of a Cognitive City with lots of subsystems and -contexts, stakeholders, interest groups and individuals this will not work. It is paramount to address explicitly the end-user in concepts for designing Cognitive Cities as this aspect touches one of its main promises: the involvement of the citizen in shaping the city, and it is necessary to sharpen the ADR-concept as it is to be expected that different organizational contexts need different ADR-concepts.

2. The artefact: Traditionally, artefacts are thought to be relatively passive in the process of development and design. However, as mentioned above, recent theories of design that were influenced by the Actor-Network-Theory suggest taking into account the impact of inanimate artefacts as non-human actors, too. This idea helps to conceptualize the role of artificial intelligence in human-centered systems like the Cognitive City: The Cognitive City-concept refers to the reciprocity of communication resp. interaction between city-related ICT and the citizens and can encompass all aspects of life in the city. The instrument to implement this idea are self-learning cognitive systems. Here, the artefact supposed to develop action-impulses by its own by collecting information and processing them independently to new knowledge. Regarding the Cognitive City, it is necessary to include in ODR the idea that artefacts are not only designed in an iterative process, but that they also have a creative effect in said process—and vice versa.

3 Designing Citizen Communication

3.1 Computational Intelligence and Citizen Communication

In the present volume the reader will find a number of contributions, both theoretical and practical, toward the successful building of a Cognitive City using techniques and methodologies from computational intelligence. The technical and theoretical aspects of such research are paramount toward the objective of a Cognitive City, but as important is the human dialogue factor: for the foreseeable future, and up to the eventual singularity (Tabacchi 2013), humans will continue to dialogically negotiate many of the aspects related to social life, in cities and elsewhere, using written and oral language and expecting as well to interact with the same means in the public discourse. As such, one important element of the development of cognitive cities will be related to automatic speech and written language recognition, on the models with which such semantization will represent the respective ontologies, as well as on the constant connection and correlation between the data captured from ubiquitous sensors and the legitimate requests for participation in the decisional process expressed through natural language. All these processes go under the name of citizen communication (Perticone and Tabacchi 2016; D'Asaro et al. 2017), and in our opinion they will be fundamental in the attainment of the main goals of all the projects in the general area of Cognitive Cities.

The ability of cognitive systems to deal with verbal and non-verbal forms of human communication is essential in the Cognitive City. Building the 'smart' components of a Cognitive City, e.g. the network of sensors, actuators and the communication infrastructure that presides to them, requires a number of advancements in electronics, miniaturization and big data handling. The problem of dealing with human communication, however, pertains more to Artificial Intelligence, and especially with methods that can handle naturally flowing information. Speech and writing recognition, in order to interface with any method of communication chosen by the citizen will be of paramount importance to bypass other, less natural input methods; linguistic register recognition, aiming at being able to interpret the non-verbal part of the discourse according to the tone and rhythm of conversation; ontology dynamical creation and update, in order to have a constant and accurate description of 'the world out there' that could be employed to direct information to the relevant parties; the informal use of formal structures (i.e. argumentation), at the aim of better understanding the kind of problems that elicit communication and to identify the actors that should play a role in the resulting conflict.

All of these tasks have in common that, as in any context where a natural language takes a predominant part, it is necessary to deal with incomplete, imprecise and missing information, uncertainty, heavy dependence from the context, an exact solution is not necessary or required (as sometimes it is even difficult to exactly pinpoint the originating problem). At the moment many of such kind of task are usually tackled by the use of a mix of brute force, statistical analysis and big data. While such approach has shown success in some research topics (speech recognition, personal digital assistants) and promising results in others (the first approaches to autonomous drive comes to mind), there are at least two obstacles to its widespread adoption. One of practical nature: some tasks, such as the understanding of the intricacies, anomalies and indeterminisms of human language, still seem out of reach such. The other, probably more important, goes at the nucleus of the problem. By using statistics and big data, introspection becomes difficult, if not impossible. And this fact is not a detail, as more often than not one of the principal reasons for AI application is not just the 'solution' of a problem by itself, but also a reflection on the way humans (and nature) tackle, solve and internalize problems. If we want to approach the problem of Citizen Communication toward a development of Cognitive Cities with the human factor in mind, we also have to look elsewhere: to Computational Intelligence.

Computational Intelligence (CI) is an umbrella term coined in the 00's to regroup all the methodologies and techniques that try to solve problems in contexts of ambiguous, incomplete, missing or vague information using approaches that are often derived from 'natural' methods, such as the ones devised by human minds or evolved in nature from animal behavior (Kacprzyk and Pedrycz 2015). In such context, algorithms derived from classical logic may be able to give an exact solution, but their requirements in time or space may be unfeasible for present and future technology, even more when the solution required has to be found in severe time constraint, such as in dynamically changing environments. CI methods are generally aimed at sub-optimal solutions that can nonetheless be achieved in a reasonable time-frame, and that are "good enough" for the intended problem (Seising 2010, 2012). A number of CI methods are directly inspired by human reasoning, especially among the earliest instances (such as Fuzzy Logic, Soft Computing and the likes), and as such are naturally matched with cognition and ideal to afford IA problems, insofar they distance themselves by grammars and inference rules.

3.2 Some Techniques

A number of articles in this book have been devoted by the authors to foundations of fuzziness, the founding father of CI (Seising and Tabacchi 2013; Termini and Tabacchi 2014; Tabacchi and Termini 2014). In this context, however, we deem appropriate to highlight some of the techniques that can be usefully applied to the problem of Citizen Communication for Cognitive Cities: Metaheuristics, Computing with Words, Computational Intelligence Classifiers, and Fuzzy-based Ontologies. The aim is not to give a detailed technical exposition of such methods, neither to review all the applicable technologies, but to give the reader the gist of the ways in which CI can be employed to implement Citizen Communication, and why this will be paramount for the functional development of Cognitive Cities thought by and for humans. Other techniques from Computational Intelligence may be usefully employed in designing Cognitive Cities (one glaring example is Fuzzy Cognitive Maps). In the rest of this volume some of such methods will be more thoroughly theoretically discussed and implemented.

3.2.1 Metaheuristics

Metaheuristics is an umbrella term designed to cover a group of 'smart' strategies to sub-optimally solve (within a certain, accepted degree of perfection) problems that are by their same nature intractable. This includes both general problems that are unsolvable due to their complexity increasing exponentially at the increase of the problem size (the so-called NP problems of the theory of computation), as well as optimization problems for which the difficulty lies in the inherent uncertainty, incomplete or imperfect information. Metaheuristics are based on the observation of ways in which nature (Evolutionary Computation, Ant colonies, Particle Swarms, and Genetic Algorithms) or humans (Simulated Annealing, Tabu Search, Local Search, Variable Neighborhood Search) solve in satisfactory ways problems that are apparently too complex to grasp, and are generally based on two assumptions: the first is that while a 'perfect' (in the computational, mathematical sense of the word) solution may not be attained, it is usually possible to find an alternative solution that satisfies the constraints put on the problem, and that is sufficiently similar, both in terms of time and space, to be acceptable. The second is that the usual way of solving a problem algorithmically—devise a number of steps that solve the problem, demonstrate the correctness, improve and repeat—can be replaced with a much more ecological procedure that explores the space of the problem in search of solution, often helped by the power of big data and high computational availability that characterizes today's computing. This exploration is often helped by sampling subsets of the problem, the use of casual choices and continuous trial and error.

There is clearly a vast space for metaheuristics in designing cognitive cities: such methods echo the way humans reason, and by mimicking human and natural strategies in solving problems they both value practical requirements over theoretical aspects, as well as promoting exploration, multiple tries, trial and error, collaboration and cooperation, further refinements as means to succeed. Furthermore, the ample availability of computing power and of graphical tools that help both experts and novice to implement metaheuristics, as well as the innate nature of such techniques for being represented as visual abstractions, may represent a way of including in the design of cognitive cities also experts from the humanities, that are usually wary of classical computation techniques.

Metaheuristics have been recently used to improve urban transportation (Nha et al. 2012): a variation of the classical Travelling Salesman problem (finding shortest routes between a number of points in a city) is extended to a number of vehicles, and a dynamical situation where traffic conditions rapidly change. One advantage of metaheuristics is that by selectively choosen subsets of the problems they can make sense of big data, especially when dynamically captured through sensors, e.g. geographic information (Cosido et al. 2013). A review of methods from agent and multiagent systems, another offshoot of metaheuristics, shows that they have been applied to many aspects of traffic and transportation systems in dynamic changing environments (Chen and Cheng 2010), as well as planning the adoption of strategies to promote sustainability in cognitive cities (Juan et al. 2011).

3.2.2 Fuzzy Sets and Fuzzy Logic

The electrical engineer and professor at the University of California Berkeley Lotfi A. Zadeh (1921–2017) used to say, "Everything is a matter of degree!" (Fig. 3). This is the insight behind his introduction of the linguistic approach to fuzzy sets and systems, and, to appreciate it, we simply have to acknowledge that we name "everything" by words. In his editorial to the first issue of the International Journal of Fuzzy Sets and Systems, he wrote that "it has become increasingly clear" that "classical mathematics—based as it is on set theory and two-valued logic—is much too restrictive and much too rigid to serve as an effective tool for the understanding

Fig. 3 Lotfi A. Zadeh giving an interview in his office, Soda Hall, University of California at Berkeley, summer of 2012 (photocredit: Fuzzy archive Rudolf Seising)

of the behavior of humanistic systems, that is, systems in which human judgment, perceptions and emotions play an important role" (Zadeh 1978).

In order to provide a mathematically exact expression of experimental research with real systems, it was necessary to employ meticulous case differentiations, differentiated terminology and definitions that were adapted to the actual circumstances, things for which the language normally used in mathematics could not account. The circumstances observed in reality could no longer simply be described using the available mathematical means. Therefore, in the summer of 1964, Zadeh was thinking about pattern recognition problems and grades of membership of an object to be an element of a class as he returned to mind almost 50 years later (Zadeh 2011; Seising 2007). Zadeh submitted his seminal article "Fuzzy Sets" to the journal Information and Control and it appeared in June 1965 (Zadeh 1965).

He introduced new mathematical entities as classes or sets that "are not classes or sets in the usual sense of these terms, since they do not dichotomize all objects into those that belong to the class and those that do not" (Zadeh 1965). He introduced "the concept of a fuzzy set, that is a class in which there may be a continuous infinity of grades of membership, with the grade of membership of an object x in a fuzzy set A represented by a number $f_A(x)$ in the interval [0,1]." He generalized the concepts, union of sets, intersection of sets, etc. He defined equality, containment, complementation, intersection and union (Fig. 4) relating to fuzzy sets A, B in any universe of discourse X as follows (for all $x \in X$):

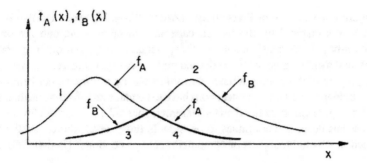

Fig. 4 Zadeh's Illustration of Fuzzy Sets in R^1: "The membership function of the union is comprised of curve segments 1 and 2; that of the intersection is comprised of segments 3 and 4 (heavy lines)" (Zadeh 1965)

- $A = B$ if and only if $\mu_A(x) = \mu_B(x)$,
- $A \subseteq B$ if and only if $\mu_A(x) \leq \mu_B(x)$,
- $\neg A$ is the complement of A, if and only if $\mu_{\neg A}(x) = 1 - \mu_A(x)$,
- $A \cup B$ if and only if $\mu_{A \cup B}(x) = \max(\mu_A(x), \mu_B(x))$,
- $A \cap B$ if and only if $\mu_{A \cap B}(x) = \min(\mu_A(x), \mu_B(x))$.

At Berkeley, Zadeh's efforts to use his fuzzy sets in linguistics led, in the early 1970s, to an interdisciplinary exchange between him and the linguist George Lakoff. The latter, referring to the accepted opinion "that sentences of natural languages (at least declarative sentences) are either true or false or, at worst, lack a truth value, or have a third value", argued "that natural language concepts have vague boundaries and fuzzy hedges and that, consequently, natural language sentences will very often be neither true, nor false, nor nonsensical, but rather true to a certain extent and false to a certain extent, true in certain respects and false in other respects" (Lakoff 1973). In this paper, Lakoff considered fuzzy sets appropriate for dealing with degrees of membership and with (concept) categories that have unsharp boundaries. Thus, Lakoff introduced the term "fuzzy logic".

Inspired and influenced by many discussions with Lakoff "concerning the meaning of hedges and their interpretation in terms of fuzzy sets", Zadeh contemplated "linguistic operators", which he called "hedges": "A basic idea suggested in this paper is that a linguistic hedge such as "very", "more", "more or less", "much", "essentially", "slightly" etc. may be viewed as an operator which acts on the fuzzy set representing the meaning of its operand."

However, based on his later research, Lakoff came to the conclusion that fuzzy logic is not an appropriate logic for linguistics: "It doesn't work for real natural languages; in traditional computer systems it works that way", he said years later. For Zadeh, fuzzy logic was the basis for "computing with words" instead of "computing with numbers" (Zadeh 1999). Later he said "the main contribution of fuzzy logic is a methodology for computing with words. No other methodology serves this purpose" (Zadeh 1996). For the new millennium, he proposed "A New Direction in AI. Toward a Computational Theory of Perceptions" (Zadeh 2001). He explained that this new approach was inspired by the human capability to operate on, and reason with, perception-based information, such as time, distance, form, and other attributes of physical and mental objects: "Everyday examples ... are parking a car, driving in city traffic, playing golf, cooking a meal, and summarizing a story. In performing such tasks ... humans base whatever decisions have to be made on information that, for the most part, is perception, rather than measurement, based." He assumed, however, that "progress has been, and continues to be, slow in those areas where a methodology is needed in which the objects of computation are perceptions" (Zadeh 2001).

3.2.3 Computing with Words

As we have discussed in the previous section, most activities in cognitive cities must be intermediated by language in order to be effective. As well, it is undenieble that a lot of the computation going on in human brains is somehow directly connected with the use of language. Computing With Words (CWW) is a recent and innovative development of Fuzziness whose aim is to replace the intermediate step of symbolic logic to offer a perspective of computation that is more directly linked to the way humans do their internal reasoning: in the words of its instigator Lotfi Zadeh "the role model for CWW is the human mind" (Zadeh 1996).

CWW offers a fusion between natural language, and specifically its verbal characteristics, and computation by standard fuzzy variables. Basic information manipulated in CWW is a collection of propositions expressed directly in words from a natural language, as a human would do. This is in contrast with classical computation, which is based on a three-steps model of transforming input into numbers or logic predicates, operating on such abstract objects and then transforming back the output of computation on something meaningful for humans. Conclusions are as such derived from premises by a double-translation procedure that introduces unwanted problems and oversimplify concepts. The principles of CWW subvert this well-established models, and put back words where they belong: as the main tool for reasoning and computing. Again, with Zadeh (1996), "in coming years, computing with words is likely to emerge as a major field in its own right. In a reversal of long-standing attitudes, the use of words in place of numbers is destined to gain respectability. This is certain to happen because it is becoming abundantly clear that in dealing with real-world problems there is much to be gained by exploiting the tolerance for imprecision. In the final analysis, it is the exploitation of the tolerance for imprecision that is the prime motivation for CWW." In CWW our basic units are called granules, and are fuzzy constraints of a variable, describing premises, some fuzzy constraint propagation, and conclusions. From the premises answers to a query expressed in a natural language are to be inferred. Computation in this instance consists in deriving such conclusions starting by the premises and using propagation.

In a Cognitive City one of the serious issues is to connect huge volumes of data that are continually and dynamically produced by ubiquitous sensors with something that can be approached by humans using natural language. The original work of systematization by symbolification by Turing (1936), extended by Chomsky (2002) and others to include language in the derivation of rules, has been hampered by inherent characteristics of the language itself: imprecision, incomplete information, uncertainty and the flexible needs a natural language has to have in its structure in order to be able to express nuances, contradictions and intended meanings (Herrera et al. 2009). CWW offers to overcome such limitation by reintroducing language as a primary citizen in computation.

The elegance and simplicity of CWW approach shows that powerful techniques need to be disruptive and to challenge the 'received' way of doing things, as the (albeit few) applications demonstrate. CWW has been employed to give a more humanlike description of phenomena linked to quality of air and pollution classification, in order to reduce respiratory problems and improve quality of life in a Cognitive City scenario (Yadav et al. 2014), as well as in order to analyze natural language expressions of symptoms and diagnoses in the assessment of ailments and other medical conditions, to the betterment of patient-doctor communication (Becker 2001). Another application of CWW is in the field of Decision Support Systems, as Decision Making is by itself defined by the natural language used to express both the rules derived by experts and the queries that are formulated in order to interrogate the system. CWW allows for a framework that bypasses the need for a hard arithmetization of rules and improves the system by making it more direct and near to the human way of dealing with decisions. A survey is to be found in (Martinez et al. 2010).

3.2.4 Computational Intelligence Classifiers

Classification is one of the main tasks in Artificial Intelligence: it helps navigation in a complex word by putting order to the chaos of our perceptions and is fundamental for a systematization of the free-flowing information captured by all kind of sensors. In the context of cognitive cities, Computational Intelligence Classifiers, and especially Fuzzy Classifiers and Fuzzy Rule-Based Classifiers add to the standard classification model the introspection abilities that are often missing from more conventional models, both in rule specification and in result analysis. With Fuzzy Classifiers, all the rules gathered from experts, mathematical models and acquisition systems alike are mediated by linguistic variables, and as such are easier to define than usual if-then rules, and more modifiable in time. At the same time, the results are built under the guise of an interpretable, linguistic base model, which can be introspected and interpreted at a higher level WRT other more common classifiers. This adds a semantic interpretation layer to the results and allows a finer tuning of parameters as well as further use of the modes obtained to be forwarded as inputs.

As in any system dealing with cognitive aspects of reality, imprecision, uncertainty and incomplete information are connatural to classification problems for cognitive cities, and Fuzzy Classifiers, have been successfully applied to such problems in the multitude. Some examples include: a Fuzzy Rule-Based System that manages the complexity of financial handling, one of the most important tasks of city government, and public resource management (Altunkaynak et al. 2005); Fuzzy Classifiers in economic planning; a Fuzzy c-means classifier for the training of a visual sensor network able to detect if parking spots are effectively free of occupied, useful as part of a public car sharing system (Baroffio et al. 2015); along with other Computational Intelligence algorithms, Fuzzy Classifiers performance have been compared in its application to cybersecurity in smart grids (Wang et al. 2010). In the health sector, a number of classifiers, including Random Forest, have been used to determine the kind of activities carried on by elderly patients by analyzing a network of smart sensors in a connected home (Nef et al. 2015), while a Fuzzy Expert System based on Fuzzy Classifiers has been employed to assess the degree of Parkinson's Disease in patients (Geman et al. 2013).

3.2.5 Fuzzy-Based Ontologies

If a Cognitive City is based, apart from sensors and actuators, on the ability to reason on what is happening and to react, as well as on a linguistic interface to communicate and drive the change, it needs the implementation of a system that allows to save and retrieve information about the state of the word, plus a facilitator for connecting and reasoning on such information in a natural fashion, such as an ontology Traditional ontologies are limited by the fact that the linkage between two entities is usually of a binary form: e.g. either a car is a vehicle or not. Using this kind of representation, a lot of subtle nuances get lost in the process, and even the resulting language becomes poorer and less expressive, having to renounce to all the quantifiers that characterize imprecise and uncertain situations. In Fuzzy Ontologies, relationships are represented by fuzzy quantities expressed in natural languages, which leads to a better representation of a complex reality, and to entities that are clearer to read and understand and can be easily manipulated to incorporate updates in a timely manner. Fuzzy ontologies are not only ideal for knowledge storing and retrieval, they are also significant in Theory of Concepts, as they supply a compatibility layer in the framework of an intentional approach. All these characteristics make the use of fuzzy ontologies suited to represent the information flowing in Cognitive Cities, as a number of assumptions about the unknown parts of such flow of data and information can be projected onto an ontology that implicitly deals with this kind of uncertainty.

Applications in which fuzzy ontologies are employed in the domain of collective communication and development include weather forecast, an environment where there are inevitable uncertainties and inconsistency associated with knowledge of natural phenomena (Truong et al. 2011), contexts that use Similarity Reasoning, for example in crisis and emergency management scenarios and other social interactions (Portmann et al. 2014). In this context the idea of Fuzzy cognitive maps, which is another fuzzy methodology with roots in cognition, should be mentioned for its contributions to the handling of knowledge (see e.g. Kaltenrieder et al. 2015). In medical domain the construction of fuzzy ontologies can support health professionals for identify useful documents in vast source of knowledge because the increasing volume of available information makes it difficult find appropriate data in different corpora pertaining to patients' healthcare needs at the time of the clinical point of view of care.

4 About the Book

This introduction aimed to give an impression of some basic ideas on designing cognitive cities which determined the conception of this book. Firstly, we introduced the terms "City", "Smart City" and "Cognitive City". Secondly, we gave an overview of design theories and approaches in order to deduce from a theoretical point of view some of the principles that needs to be taken into account when designing the Cognitive City. The third part highlighted some concrete techniques that can be usefully applied to the problem of citizen communication for Cognitive Cities. We want to conclude with a short overview of the following nine chapters.

Next to this introductory section the book consists of two parts. The first five articles deal with conceptual considerations. Alejandro Sobrino ("*Cognitive Cities: An Approach from Citizens*") explores the role of the body and the brain in human cognition and its relevance for Cognitive Cities. To this end the author discusses the role of smart textiles to transform the citizens into kinds of transducers between external stimuli and their cognition. The author proposes cognitive maps as a skeletal method for explaining concepts and for detecting typical social behaviors, such as cooperation, competition or mediation. Miroslav Hudec ("*Possibilities for Linguistic Summaries in Cognitive Cities*") points out that citizens wish to be part of decision-making processes and to be informed about various developments in cities. Linguistic summaries based on the fuzzy sets and fuzzy logic theory could be the solution to keep the stakeholders informed concisely, as linguistic summaries are able to verbalize mined information from the data by quantified sentences of natural language.

Enric Trillas, Sara D'Onofrio and Edy Portmann ("*An Exploration of Creative Reasoning*") introduce a model of creative reasoning and give first hints of a possible mathematical model to enable this. The authors emphasize the importance of creativity in human life. A mathematical structure that allows to include 'creative jumps' in reasoning is elaborated. Sara D'Onofrio, Elpiniki Papageorgio and Edy Portmann ("*Using Fuzzy Cognitive Maps to Arouse Learning Processes in Cities*") investigate how Fuzzy cognitive maps (FCMs) can be used to model interconnected and imprecise urban data which are usually expressed in natural language and thus imprecise but can contain relevant information that should be processed to advance the city. As cognitive learning processes are crucial to enhance the smartness of a city this can be optimized by combining FCMs with learning algorithms. José M. Alonso, Ciro Castiello and Corrado Mencar ("*The Role of Interpretable Fuzzy Systems in Designing Cognitive Cities*") observe that researchers and developers created a novel generation of intelligent systems which are producing intelligent devices, yielding what is called smart cities. Fuzzy systems are already used in this context of Smart Cities. To pass from smart to cognitive cities the next step is to address the interaction between intelligent systems and citizens which can be facilitated using interpretable fuzzy systems.

The following four chapters deal with concrete applications of cognitive city-concepts. Javier Cuenca, Felix Larrinaga, Luka Eciolaza and Edward Curry (*"Towards Cognitive Cities in the Energy Domain"*) highlight the importance of the Semantic Web and semantic ontologies as a foundation for learning and cognitive systems. They show that the energy sector is a potential field of action to apply the Cognitive City-approach. Mo Mansouri and Nasrin Khansari (*"A Conceptual Model for Intelligent Urban Governance: Influencing Energy Behavior in Cognitive Cities"*) explore by the example of energy consumption the influence of governance on sustainability. They map the impact of information and communication technologies (ICTs) on the decision-making process by increasing policy effectiveness, accountability, and transparency within urban systems.

Kurt Bollacker, Natalia Díaz Rodríguez and Xian Li (*"Extending Knowledge Graphs with Subjective Influence Networks for Personalized Fashion"*) introduce the Stitch Fix's industry case as an applied fashion application in cognitive cities. The authors explain that it is still a challenge to extract knowledge and actionable insights from fashion data due to the intrinsic subjectivity needed to effectively model the domain. They address this by proposing a supplementary ontological approach. On the basis of a set of use cases possible classes of prediction questions and machine learning experiments are discussed as well as a case study on business models and monetization strategies for digital fashion. Finally, Patrick Kaltenrieder, Jorge Parra, Thomas Krebs, Noémie Zurlinden, Edy Portmann and Thomas Myrach (*"A Dynamic Route Planning Prototype for Cognitive Cities"*) present a software prototype for dynamic route planning in cognitive cities. In contrast to already existing tools, the prototype gives the user additional flexibility, which is especially interesting in the tourist sector.

References

Albino V, Berardi U, Dangelico RM (2015) Smart cities: definitions, dimensions, performance, and initiatives. J Urban Technol 22(1):3–21

Altunkaynak A, Özger M, Çakmakci M (2005) Water consumption prediction of Istanbul city by using fuzzy logic approach. Water Resour Manag 19(5):641–654

Baroffio L, Bondi L, Cesana M, Redondi AE, Tagliasacchi M (2015) A visual sensor network for parking lot occupancy detection in smart cities. In: 2015 IEEE 2nd world forum on internet of things (WF-IoT), pp 745–750

Becker H (2001) Computing with words and machine learning in medical diagnostics. Inf Sci 134(1–4):53–69

Beinrott V (2015) Bürgerorientierte Smart City: Potentiale und Herausforderungen. The Open Government Institute, Friedrichshafen

Bergson H (1907) L'Evolution créatrice. Aclan, Paris

Caine R, Caine G (2011) Natural learning for a connected world: education, technology and the human brain. Teachers College Press, New York

Chen B, Cheng HH (2010) A review of the applications of agent technology in traffic and transportation systems. IEEE Trans Intell Transp Syst 11(2):485–497

Chomsky N (2002) Syntactic structures. Walter de Gruyter, Berlin

Cohen B (2015) The 3 Generations of Smart Cities: Inside the development of the technology driven city. Fastcompany. https://www.fastcompany.com/3047795/the-3-generations-of-smart-c ities. Accessed 2 Mar 2018

Cosido O, Loucera C, Iglesias A (2013) Automatic calculation of bicycle routes by combining meta- heuristics and gis techniques within the framework of smart cities. In: 2013 International conference on new concepts in smart cities: fostering public and private alliances (SmartMILE). IEEE, pp 1–6

D'Asaro FA, Di Gangi MA, Perticone V, Tabacchi ME (2017) Computational intelligence and citizen communication in the smart city. Inform-Spektrum 40(1):25–34

Dyer M, Corsini F, Certomà C (2017) Making urban design a public participatory goal: toward evidence-based urbanism. Proc Inst Civ Eng Urban Des Plann 170(2):173–186

Eckardt F (2014) Stadtforschung, Gegenstand und Methoden. Springer, Wiesbaden

Finger M, Portmann E (2015) Smart Cities—Ein Überblick. HMD Praxis der Wirtschaftsinformatik 52(4):470–481

Finger M, Portmann E (2016) What are cognitive cities? In: Finger M, Portmann E (eds) Towards cognitive cities: advances in cognitive computing and its application to the governance of large urban systems. Springer International Publishing, Heidelberg, pp 1–11

Geman O, Turcu CO, Graur A (2013) Parkinson's disease assessment using fuzzy expert system and nonlinear dynamics. Adv Electr Comput Eng 13(1):41–46

Gregor S, Jones D (2007) The Anatomy of a design theory. J Assoc Inf Syst 8(5):312–335

Herrera F, Alonso S, Chiclana F, Herrera-Viedma E (2009) Computing with words in decision making: foundations, trends and prospects. Fuzzy Optim Decis Mak 8(4):337–364

Hirdina H (2010) Design. In: Barck K et al (eds) Ästhetische Grundbegriffe, vol 2. Metzler, Stuttgart, pp 41–63

Hurwitz JS, Kaufman M, Bowles A (2015) Cognitive computing and big data analysis. Wiley, Indianapolis

Iivari J (2007) A paradigmatic analysis of information systems as a design science. Scand J Inf Syst 19(2):39–64

Ito M, Gutiérrez K, Livingstone S, Penuel B, Rhodes J, Salen K, Schor J, Sefton-Green J, Watkins SC (eds) (2013) Connected learning: an agenda for research and design. Digital Media and Learning Research Hub, Irvine

Juan YK, Wang L, Wang J, Leckie JO, Li KM (2011) A decision-support system for smarter city planning and management. IBM J Res Dev 55(1.2):3–1

Kacprzyk J, Pedrycz W (2015) Springer handbook of computational intelligence. Springer, New York

Kaltenrieder P, D'Onofrio S, Portmann E (2015) Enhancing multidirectional communication for cognitive cities. In: Second IEEE international conference eDemocracy and eGovernment (ICEDEG 2015). Quito, pp 38–43

Krippendorff K (2006) The semantic turn: a new foundation for design. Taylor & Francis, Boca Raton

Lakoff G (1973) Hedges: a study in meaning criteria and the logic of fuzzy concepts. J Philos Log 2:458–508

Maccani G, Donnellan B, Helfert M (2014a) Action design research in practice: the case of smart cities. In: Tremblay MC, VanderMeer D, Rothenberger M, Gupta A, Yoon V (eds) DESRIST 2014: advancing the impact of design science: moving from theory to practice. Lecture Notes in Computer Science, vol 8463. Springer, Cham, pp 132–147

Maccani G, Donnellan B, Helfert M (2014b) Systematic problem formulation in action design research: the case of smart cities. In: Proceedings of the European conference on information systems (ECIS) 2014, Tel Aviv, Israel, 9–11 June 2014. https://pdfs.semanticscholar.org/26b1/1 8ae91e258111cdd09e8af3a95956cd24faa.pdf. Accessed 20 Mar 2018

Malone TW, Bernstein MS (2015) Introduction. In: Malone TW, Bernstein MS (eds) Handbook of collective intelligence. MIT Press, Cambridge, MA, pp 1–13

Mareis C (2011) Design als Wissenskultur. Interferenzen zwischen Design- und Wissensdiskursen seit 1960. Transcript, Bielefeld

Mareis C (2013) Wer gestaltet die Gestaltung? Zur ambivalenten Verfassung von Partizipation und Design. In: Mareis C, Held M, Joost G (eds) Wer gestaltet die Gestaltung? Praxis, Theorie und Geschichte des Designs. Transcript, Bielefeld, pp 9–20

Mareis C (2014) Theorien des Design. Junius-Verlag, Hamburg

Martínez L, Ruan D, Herrera F (2010) Computing with words in decision support systems: an overview on models and applications. Int J Comput Intell Syst 3(4):382–395

Marx K (1959) Das Kapital, Bd. I [1867], Marx/Engels-Gesamtausgabe, 23. Dietz Verlag, Berlin

Mieg HA (2013) Einleitung. Perspektiven der Stadtforschung. In: Mieg HA, Heyl C (eds) Stadt: ein interdisziplinäres Handbuch. Metzeler, Stuttgart, pp 1–14

Nef T, Urwyler P, Büchler M, Tarnanas I, Stucki R, Cazzoli D, Müri R, Mosimann U (2015) Evaluation of three state-of-the-art classifiers for recognition of activities of daily living from smart home ambient data. Sensors 15(5):11725–11740

Nha VTN, Djahel S, Murphy J (2012) A comparative study of vehicles routing algorithms for route planning in smart cities. In: First international workshop 2012 vehicular traffic management for smart cities (VTM), pp 1–6

Orlikowski W, Iacono C (2001) Desperately seeking the "IT" in IT research—a call to theorizing the IT artifact. Inf Syst Res 12(2):121–134

Perticone V, Tabacchi ME (2016) Towards the improvement of citizen communication through computational intelligence. In: Portmann E, Finger M (eds) Towards cognitive cities. Springer, New York, pp 83–100

Portmann E, Meier A, Cudré-Mauroux P, Pedrycz W (2014) FORA—a fuzzy set based framework for online reputation management. Fuzzy Sets Syst 269:90–114

Rittel HWJ (1987) The reasoning of designers. In: International congress on planning and design theory, Boston. http://docshare01.docshare.tips/files/26150/261507822.pdf. Accessed 27 Mar 2018

Rittel HWJ (2013) Zur Planungskrise. Systemanalyse der "ersten und der zweiten Generation". In: Reuter WD, Jonas W, Rittel H (eds) Thinking design. Transdisziplinäre Konzepte für Planer und Entwerfer. Birkhäuser, Basel, pp 20–38

Ropohl G (2010) Technik, anthropologisch, oder: Homo faber, http://www.ropohl.de/4.html. Accessed 14 Feb 2018

Scheler M (1960) Die Wissensformen und die Gesellschaft [1926]. Francke, Bern

Sein MK, Henfridsson O, Purao S, Rossi M, Lindgren R (2011) Action design research. MIS Q 35(1):37–56

Seising R (2007) The fuzzification of systems: the genesis of fuzzy set theory and its initial applications – developments up to the 1970s. Springer, Berlin

Seising R (2010) What is Soft Computing? Bridging Gaps for the 21st Century Science! Int J Comput Intell Syst 3(2):160–175

Seising R (2012) Science visions, science fiction and the roots of soft computing. In: Moewes C, Nürnberger A (eds) Computational intelligence in intelligent data analysis. Essays Dedicated to Rudolf Kruse on the Occasion of his 60th Birthday. Springer, Berlin, pp 123–150

Seising R, Tabacchi ME (2013) A very brief history of soft computing: fuzzy sets, artificial neural networks, and evolutionary computation. In: Proceedings of the 2013 joint IFSA world congress NAFIPS annual meeting (IFSA/NAFIPS) edmonton, Canada, 24–28 June 2013, pp 739–744

Siemens G (2006) Knowing Knowledge. Lulu.com; http://www.elearnspace.org/KnowingKnowle dge_LowRes.pdf. Accessed 2 Mar 2018

Simon HA (1969) The Sciences of the artificial. MIT Press, Cambridge

Ståhlbröst A, Padyab A, Sällström A, Hollosi D (2015a) Design of Smart City Systems from a Privacy Perspective. IADIS Int J WWW/Internet 13(1):1–16

Ståhlbröst A, Bergvall-Kåreborn B, Ihlström-Eriksson C (2015b) Stakeholders in smart city living lab processes. In: Americas conference on information systems: 13/08/2015–15/08/2015. Americas Conference on Information Systems. https://www.diva-portal.org/smash/get/diva2:100602 7/FULLTEXT01.pdf. Accessed 20 Mar 2018

Tabacchi ME (2013) Salvi ed al sicuro: Singolarità di Vinge, crescita tecnologica, limiti di processo. In: Auricchio A, Cruciani M, Rega M, Villani M (eds) Scienze cognitive: paradigmi sull'uomo e la tecnologia. Atti del X convegno annuale AISC (Giornale italiano di Neuroscienze, Psicologia e Riabilitazione 2), pp 213–218

Tabacchi ME, Termini S (2014) Some reflections on fuzzy set theory as an experimental science. In: Information processing and management of uncertainty in knowledge-based systems, vol. 442 of Communications in Computer and Information

Termini S, Tabacchi ME (2014) Fuzzy set theory as a methodological bridge between hard science and humanities. Int J Intell Syst 29(1):104–117

Townsend AM (2013) Smart cities: big data, civic hackers, and the quest for a New Utopia. W.W. Norton Inc., New York

Truong HB, Nguyen NT, Nguyen PK (2011) Fuzzy ontology building and integration for fuzzy inference systems in weather forecast domain. Intelligent information and database systems. Springer, Berlin, pp 517–527

Turing A (1936) On computable numbers, with an application to the Entscheidungsproblem. J Math 42(1):230–265

United Nations, Department of Economic and Social Affairs, Population Division (2014) World urbanization prospects: The 2014 revision, highlights (ST/ESA/SER.A/352), http://esa.un.org/u npd/wup/Publications/Files/WUP2014-Highlights.pdf. Accessed 12 Mar 2018

United Nations, Department of Economic and Social Affairs, Population Division: World Population Prospects (2015) The 2015 revision, key findings and advance tables. Working paper no. ESA/P/WP.241, http://esa.un.org/unpd/wpp/Publications/Files/Key_Findings_WPP_2015. pdf. Accessed 12 Mar 2018

Vaishnavi V, Kuechler W (2015) Design science research in information systems, January 20, 2004; Last updated: 15 Nov 2015, http://www.desrist.org/design-research-in-information-sys tems. Accessed 12 Mar 2018

Walker JA (1989) Design history and the history of design. Pluto, London

Wang Y, Ruan D, Xu J, Wen M, Deng L (2010) Computational intelligence algorithms analysis for smart grid cyber security. In: International Conference in Swarm Intelligence, pp 77–84

Wilke G, Portmann E (2016) Granular computing as a basis of human-data interaction: a cognitive cities use case. Granul Comput 1:181–197

Willis AM (2006) Ontological designing. Des Philos Pap 4(2):69–92

Yadav J, Kharat V, Deshpande A (2014) Fuzzy description of air quality using fuzzy inference system with degree of match via computing with words: a case study. Air Qual Atmos Health 7(3):325–334

Zadeh LA (1965) Fuzzy Sets. Inf Control 8:338–353

Zadeh LA (1978) An editorial perspective. Int J Fuzzy Sets Syst 1(1):1

Zadeh LA (1996) Fuzzy logic = computing with words. IEEE Trans Fuzzy Syst 4(2):103–111

Zadeh LA (1999) From computing with numbers to computing with words: from manipulation of measurements to manipulation of perceptions. IEEE Trans Circuits Syst I Fundam Theory Appl 46(1):105–119

Zadeh LA (2001) A new direction in AI: toward a computational theory of perceptions. AI-Mag 22(1):73–84

Zadeh LA (2011) My life and work–a retrospective view. Appl Comput Math 10(1):4–9

Marco E. Tabacchi is currently the Scientific Director at Istituto Nazionale di Ricerche Demopolis, Italy. He is a transdisciplinary scientist, with interests in cognitive science, computational intelligence applied to IoT and philosophy and history of soft computing.

Edy Portmann is a researcher and scholar, specialist and consultant for semantic search, social media, and soft computing. Currently, he works as a Swiss Post-Funded Professor of Computer Science at the Human-IST Institute of the University of Fribourg, Switzerland. Edy Portmann studied for a BSc in Information Systems at the Lucerne University of Applied Sciences and Arts, for an MSc in Business and Economics at the University of Basel, and for a Ph.D. in Computer Sciences at the University of Fribourg. He was a Visiting Research Scholar at National University of Singapore (NUS), Postdoctoral Researcher at University of California at Berkeley, USA, and Assistant Professor at the University of Bern. Next to his studies, Edy Portmann worked several years in a number of organizations in study-related disciplines. Among others, he worked as Supervisor at Link Market Research Institute, as Contract Manager for Swisscom Mobile, as Business Analyst for PwC, as IT Auditor at Ernst & Young and, in addition to his doctoral studies, as Researcher at the Lucerne University of Applied Science and Arts. Edy Portmann is repeated nominee for Marquis Who's Who, selected member of the Heidelberg Laureate Forum, co-founder of Mediamatics, and co-editor of the Springer Series 'Fuzzy Management Methods', as well as author of several popular books in his field. He lives happily married in Bern and has three lively kids.

Rudolf Seising obtained his Ph.D. in Philosophy of Science and his Habilitation in History of Science from the Ludwig-Maximilians-University in Munich, Germany, after studies of Mathematics, Physics and Philosophy at the Ruhr-University Bochum, Germany. Rudolf Seising has been Scientific Assistant for Computer Sciences (1988–1995) and for History of Sciences (1995–2002) at the University of the Armed Forces in Munich. From 2002 to 2008 he was with the Core unit for Medical Statistics and Informatics at the University of Vienna Medical School, Austria. He acted as Professor for History of Science at the Friedrich-Schiller-University Jena, Germany, and at the Ludwig-Maximilians-University in Munich. He works now at the Research Institute for the History of Science and Technology at the Deutsches Museum in Munich and is Lecturer at the Faculty of History and Arts at the Ludwig-Maximilians-University Munich. Rudolf Seising was Visiting Researcher (2008–2010) and Adjoint Researcher (2010–2014) at the European Centre for Soft Computing in Mieres (Asturias), Spain, and he has been several times Visiting Scholar at the University of California in Berkeley, USA. He is Chairman of the IFSA Special Interest Group "History" and of the EUSFLAT Working Group "Philosophical Foundations". He is member of the IEEE Computational Intelligence Society (CIS) History Committee and of the IEEE CIS Fuzzy Technical Committee. His main areas of research comprise the historical and philosophical foundations of science and technology.

Astrid Habenstein leads the Transdisciplinary Research Centre Smart Swiss Capital Region (TRCSSCR) at the University of Bern, Switzerland. She graduated at the University of Bielefeld, Germany, in History and Philosophy (2005) and received her Ph.D. in Ancient History at the University of Bern (2012). Astrid Habenstein worked as Research Assistant at the University of Bern and as Lecturer at the Universities of Bern and Basel. She was member of Edy Portmanns project on Smart and Cognitive Cities at the Institute of Information Systems, University of Bern (2016–2017). Her main interdisciplinary and transdisciplinary research interests include Smart and Cognitive Cities, sociological theory formation (knowledge research, theories of design, systems theory, theories of interaction) and research development.

Part II
Concepts

Cognitive Cities: An Approach from Citizens

Alejandro Sobrino

Abstract Cities are made from both physical entities (buildings and its links) and social entities (citizens and its links). Cognitive cities demand cognitive citizens. In this paper, we advocate the role of the body and the brain in human cognition. Citizens are rational, embodied agents. As embodied beings, we claim the role of the smart textiles to transform them into kinds of transducers between external stimuli and their cognition. As rational beings, we defend explanations as a relevant way for securing successful adaptations in dynamic scenarios, such as those of cognitive cities. We propose cognitive maps as a skeletal method for explaining concepts and for detecting certain typical social behaviors, such as cooperation, competition or mediation.

Keywords Cognitive cities · Embodied cognition · Smart textile · Cognitive maps Social behavior

1 Introduction

Cities arose in the Neolithic, a period during which the passage from nomadism to the sedentary life style took place, and animals and plants were domesticated, giving rise to agriculture and the raising of livestock. At that time, cities emerged from the need to control surplus production and as a way for people to defend themselves against potential thieves. Thus, O'Flaherty (2005) placed the origin of cities in military protection, initially in charge of groups of inhabitants and later professionalized with the emergence of the army.

From the origin, groupings of people have not stopped growing. Today, approximately half of the world's population lives in urban areas. Urban density reduces the costs of basic demands, such as transport, education, security or health, favoring the success of cities.

A. Sobrino (✉)
University of Santiago de Compostela, Santiago de Compostela, Spain
e-mail: alejandro.sobrino@usc.es

© Springer Nature Switzerland AG 2019
E. Portmann et al. (eds.), *Designing Cognitive Cities*, Studies in Systems, Decision and Control 176, https://doi.org/10.1007/978-3-030-00317-3_2

The city can be characterized by considering at least three different entities: (a) the people who inhabit it, (b) the services it provides and (c) the law and norms that regulate the activity of its citizens.

From the point of view of population, a city is a permanent grouping of people characterized by a significant density in a large geographic area. Tokyo is currently the most densely populated and largest city in the world.

From the perspective of the services it provides, the city facilitates the interaction of people with a view to exchanging tangible and intangible goods: salary for work; taxes for health or leisure for money, to give but a few examples. To satisfy the citizens' needs, industrial parks, specialized hospital areas or large entertainment centers are built.

From the view of the administration, the city has a legal and administrative frame-work that provides the local law by which its inhabitants are governed. People who are registered in the city are called citizens (of that city) and as such, have rights and must observe certain duties.

All citizens are cognitive agents, but specifying what cognition is becomes a complex task. Prima facie, cognition is a consequence of using our senses to represent the world, to categorize it and to transmit our abstractions to others. Thus, the phrase 'This boy has cognitive deficiencies' probably suggests that his or her perception or categorization of the world is anomalous. Nonetheless, cognition can also be understood as the very process by which we access to the knowledge of something. Thus, the sentence. 'The confusion comes from several cognitive mistakes' presumably denotes that the representation chain of the world is not adhered to correctly, or is supplanted by an erroneous one. In sum, cognition can be understood as the end of a sequence by way of which we know something, or as the very process of accessing that knowledge (Cfr. Brandimonte et al. 2006). Cognition as a fact is denoted by a name (cognition, knowledge), but cognition as a process by a verb (to recognize, to know). Both grammar categories remit to different philosophical traditions: the analytic and the phenomenological ones.

Analytical philosophers advocated for objective knowledge. To know is to find a mapping between language (or another representational system) and the world. Wittgenstein's Tractatus (1914–1916) paradigmatically supported this view. Phe-nomenologists, such as Merleau-Ponty (1945), defended cognition as a process. To know is a kind of encounter between the world and a subject, to whom the world is presented. Direct knowledge is self-validating and has an epistemological preva-lence over a more elaborated or assisted knowledge, such as scientific knowledge. In the vein of Phenomenologists, Gestalt theorists, such as Koffka, Wertheimer, or Köhler, distinguished two phases in the process of representation: a qualitative one, wherein cognition is related with direct access to the object as experienced by our senses (phenomenological influence), and a rational or cognitive one, founded objec-tively (analytic ascendency) (Cfr. Epstein and Hatfield 1994). As Phenomenologists, Gestalt's supporters defended the methodological primacy of the former phase over the latter: to know is primarily for there to be an impact in our senses; later, justifica-tion takes place. In the vein of Gestalt theorists, we advocate the role of the body and the brain in cognition, allowing the body to manage stimuli and the brain to objectify them.

In this paper we will discuss the city from the perspective of cognitive citizens. Citizens actively participate in the life of cities and constitute their human fabric. They are rational agents made up of brain and body. Since Descartes, only the brain was relevant in human cognitions, leaving the body in charge of the instincts, typical of non-human animals. Nevertheless, contemporary studies show the relevance of muscles and glands in human cognition.

We now go on to outline the structure of our work. In Sect. 2 we highlight the role of the body in human cognition, the so-called embodied cognition, and its influence in the modern Artificial Intelligence, presenting the smart textile as a kind of transducer between the human body and its environment, providing a kind of data source indicative of cognitive states. In cities, the division of labor is imposed, owing to the specialization of tasks. Some people will ask others to change or modify their behavior based on an authoritative criterion. The success of that demand will largely depend on the extent to which the imposition or suggestion is explained. In Sect. 3 we show the role of vague language in the socialization and, hence, in the explanations involving citizens, addressing the role of cognitive maps explaining causal scenarios in a schematic way, facilitating the detection of social behaviors, such as cooperation, conflict or arbitration. In Sect. 4 we introduce the role of semantic vagueness and node relevance in order to better adapt the similarity or divergence between cognitive maps. In Sect. 5 we summarize the main ideas and, finally, we end with the references.

2   Embodied Cognition and Smart Textile

Although cognition depends largely on the brain, other parts of the body actively participate in it, such as the eyes, nose or feet. Embodied cognition is closely related to the embodiment thesis, according to which many characteristics of cognition are dependent on the body (including the brain but not limited to it). Thus, the body has a constitutive or a causal role in our knowledge. If constitutive, cognition is nothing but body cognition. If causal, the body helps to explain cognition. We adhere to the latter position.

It is common to attribute three causal roles to the body in embodied cognition (Clark 1997). The body may act as: (a) a restrictor on cognition; (b) a distributor of the cognitive process; and (c) a regulator of cognitive activity. We now go on to describe each one below.

The agent's body significantly restricts the nature and content of cognition. Human beings smell less than a dog, run slower than an ostrich or see less than a bee (sensitive to ultraviolet light). Our senses prevent us from seeing in a dark room, although a bats can fly around with no problems, thanks to their echolocation. In addition, some cognition is dependent more on body movements than on rational justifications. The work by Varela et al. (1992) on 'enacted knowledge' illustrates this: I ride my bicycle by practicing on it, not by defining and learning the rules governing balance.

The body also acts by distributing computational functions. It receives inputs from the outside, such as a thermal sensation, and transmits them to the brain, which, having processed them, may issue a motor command involving different muscles in a feedback process.

Finally, the agent's bodily functions regulate cognitive activity in space and time, coordinating cognition and action. For instance, a heat receptor on the skin detects a high temperature and the brain issues an order to the muscles in a very short time requiring them to move away from the heat source and, thus, avoid potential damage. Cognition is effective only if there is temporal and spatial coordination in the interpretation of spatially timed events.

In sum, it is the whole body, and not only the brain, which is causally relevant in cognition. This is also the thesis applied by Dreyfus (1986, 1992) to the field of the Artificial Intelligence (AI), pioneering the benefits of embedded AI, currently one of the dominant paradigms of this modern science. Contrary to Descartes, who underestimated the body, Dreyfus defends a cognition that is situated and embodied. We briefly explain these notions below.

Cognition is situated because our knowledge is context-dependent, often obeying local situations and intended to solve specific problems. For instance, a pen is for writing, but in a robbery scenario, it could be used as a kind of weapon against the thief. And we can imagine other uses: for rewinding a cassette, as a blowpipe or to put one's hair up in a bun. As Waismann (1945) asserts, most meanings have open texture, are not concrete, and therefore show essential vagueness. Thus, our knowledge must be also open-minded, providing new interpretations of established facts, as the history of knowledge shows. Human beings were born in, and interact with, a world that is more open than closed, more subjective than objective, and more imprecise than crisp.

Cognition is embodied because is performed by a body. Brain is a major part of the body, but it needs interfaces with the outside. Interpretations or justification of perceptions are made after the object is presented to our senses. Concerning language, Hauser et al. (2002) classified the interfaces into two categories: 'conceptual-intentional' and 'sensory-motor' ones, although only the latter is relevant for our purposes. The sensory-motor interface is responsible for perception, recognition and approximation to the object, integration and coordination. It includes skills such as pattern recognition, clustering, manual manipulation of objects, transformation of spatial structures and mental anticipation of the effects. It depends more on muscles and glands than on gray matter. To Dreyfus, sensory-motor intelligence is previous and primary to the rational one—a thesis also supported by Piaget in his approach to human cognitive development- placing the body at the center of the modern Artificial Intelligence agenda.

Influenced by Dreyfus, Artificial Intelligence (AI) has recently moved to embodied AI in an attempt to build robots that perceive their environment, adapting to solve problems (Cfr. Hoffman 2012). For that task, they are equipped with a large quantity of sensors. Embedded robots would undoubtedly be more cognitive than those simply managing data; i.e., helping people to select a route or to improve their electricity consumption.

Unlike robots, human beings are cognitive agents as we have a sensitive body and a reasoning brain. Until recently the body was simply a receiver of signals. But its cognitive power can be increased if it can be transformed into an information transducer. That is the role of the smart textile.

Wearing smart clothes, people can automatically or involuntarily send signals to other individuals or robots reporting their physical or emotional state. This may be useful for conveying recommendations or mandates. For example, if the municipality sends a fine to a neighbor and his biometric analysis shows that he or she is angry, perhaps the authorities should wait and report the ticket later. Next, we will approach the role of smart textiles in augmented body cognition.

Humans used textile garments to cover themselves and as protection from the cold and inclement weather for 30,000 years. Fabrics have played an important role in the lives of people in the form of garments or household clothes. Today, in addition to those basic functions, they also at the service of fashion, providing aesthetic wellbeing, offering a enormous array of textures, colors and shapes. In the past, garments were made of fabric with metal or ceramic inserts to facilitate their operation, such as buttons or zippers; or implementing ornaments, such as sequins or beads. In recent times, fabrics have been able to incorporate more interesting accessories, such as sensors of different types, transforming them into active devices, capable of sending or receiving signals to and from the body, acting as real information transducers. Thus, for instance, a sensor can constantly measure the body temperature and send a signal to a medical center if a threshold is exceeded or detect a contracture and automatically ask for an expert evaluation of it. Sensors can also transmit personal emotions, as the feeling of a hug while having a videoconference with a friend.

The crucial components of smart textiles are sensors and actuators, along with control units to manage their activity. Two types of intelligent textiles are distinguished: passive textiles, capable of detecting the environment (Langenhove et al. 2005), and active textiles, which contain sensors but also actuators to keep, for instance, the body temperature constant (Hildebrant et al. 2015). Sensors must be used in clothes without losing the classic properties of flexibility, comfort, durability and cleanable and fashionable appearance. To do so, silicon intelligence and its physical supports need to be integrated into a soft textile substrate. To solve this challenge, nanotechnology provides promising solutions. Integrating intelligence at the nanoscale, it is possible to preserve the mechanical properties of the fabrics to achieve smart clothes widely accepted by the general public (Coyle et al. 2007).

As mentioned above, textiles not only have a role in human clothing, but also in the house or office furniture (pillows, sofas, curtains) or even in leisure or work accessories, such as in car seats or toy kites, being able to register, send or exchange information that is useful for the user, institution or company. Smart textiles can become a universal and ubiquitous interface, monitoring the wearer and the environment and reacting to external stimuli, whether mechanical, electrical, chemical or thermal.

There are already a number of industrial projects in smart textiles. Next, I will only mention some of them. Siren Care uses temperature sensors to detect inflammation and injury in real time for sufferers of diabetes. Textronics developed NuMetrex, a heart rate monitor built-in into a bra. Philips Lumalive integrates flexible inlays of colored LEDs into the fabric, emitting light and displaying dynamic messages and graphics or multicolored images, improving the observer's mood and positively influencing their behavior. Finally, the Galician (Spain) company, SCIO TT, recently developed S2S, an intelligent maritime safety vest capable of alerting the emergency services in case of a man falling overboard. These examples show that the development of Smart Textiles can provide garments that are contextually aware and tailored to individual users' needs.

3 Cognitive Maps for Explaining Social Interaction

Smart clothes look at the external, but an innermost aspect of citizens should be considered: how they feel, how they perceive problems, and how they are convinced by experts to modify past behaviors in order to adopt new solutions in a constantly changing, dynamic environment. In this task, the justification of the actions, imposed or suggested by the authorities, acquires great relevance. The success of this task will depend, to a large extent, on the explanation provided. Explanations are mainly causal and help to understand why we must take one course of action and not another.

Today the majority of people live in society and mutual respect is essential. In modern cities, the division of labor is imposed, leading to specialization and expertise. This gives rise to specialists, who, from a position of authority, recommend the acceptance new rules or implement new routines. City legislators often integrate these suggestions consistently into the legal tradition, so that they are not contradictory with previously promulgated norms. Acceptance of these recommendations or impositions depends, to a large extent, on how they are explained.

Causal explanation is a cognitive tool for organizing the world, for justifying the actions taken thereupon and for convincing citizens of their appropriateness and rational foundation. For instance, the increase in population in a neighborhood may justify new streets, or the decrease in crime in a street may objectify a redistribution of the staff in a police station.

Many explanations are causal and help to understand why we must take one course of action and not another. In order to cognitively grasp an event, we can use written or oral language, or draw a schematic representation of it, the latter perhaps being a more intuitive and explanatory procedure than the former, as Paivio argued in his dual coding theory (Cfr. Paivio 1986). Larkin and Simon (1987) advocated for a diagrammatic representation instead of a verbal one, as the schemes present all the information at once, grouping together functionally similar elements and thus facilitating perceptual inferences, which are extremely easy for human beings. The use of one representational method or the other is, however, not exclusive, and used

Table 1 Incidence matrix corresponding to Fig. 1

	C1	C2	C3	C4	C5	C6	C7
C1	0	0	0	0	0	0	1
C2	1	0	1	0	0	0	0
C3	1	1	0	1	0	0	0
C4	1	0	1	0	0	0	0
C5	0	−1	−1	0	0	0	0
C6	0	−1	−1	0	0	0	0
C7	0	1	0	1	0	0	0

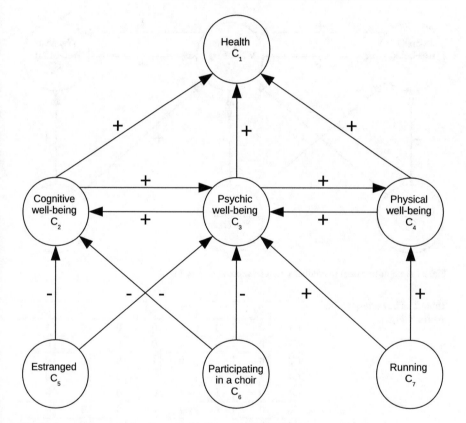

Fig. 1 A cognitive map showing causal influence in the health area

in conjunction contribute to clarify the problem. Cognitive maps are a common way for graphically explaining causal dependencies between actions or agents.

First introduced by Axelrod (1976), a cognitive map can be used to represent causal relations and to provide explanations of causal influences. A cognitive map has nodes and arcs. Nodes are labeled with concepts and arcs represent the causal

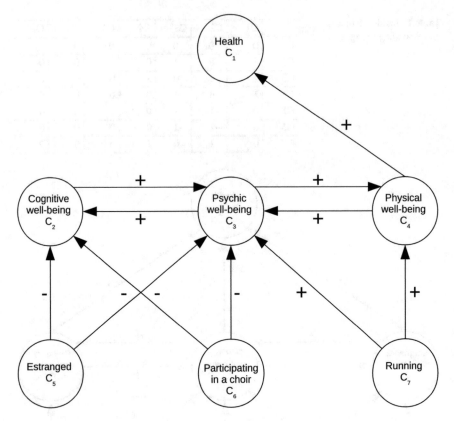

Fig. 2 A cognitive map in conflict with that represented by Fig. 1

Table 2 The corresponding matrix to Fig. 2

	C1	C2	C3	C4	C5	C6	C7
C1	0	0	0	0	0	0	1
C2	0	0	1	0	0	0	0
C3	0	1	0	1	0	0	0
C4	1	0	1	0	0	0	0
C5	0	−1	−1	0	0	0	0
C6	0	−1	−1	0	0	0	0
C7	0	1	0	1	0	0	0

influence, which may be (a) positive (favoring the effect, 1), (b) negative (inhibiting the effect, −1) or (c) neutral (no influence on the effect, 0).

The following cognitive map uses this three-valued notation to illustrate the causal influence of several factors on health.

The graph can be summarized into a matrix, where rows and columns host nodes, and the intersection hosts the causal influence. If there is a link between two nodes

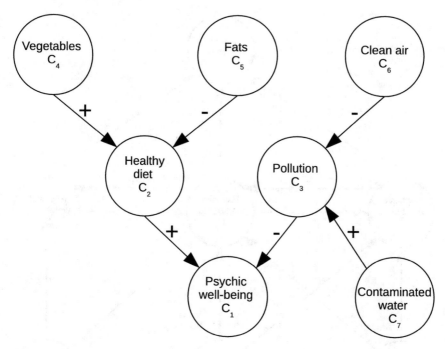

Fig. 3 Other cognitive map with factors influencing health

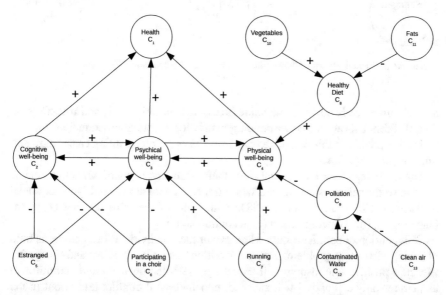

Fig. 4 A cognitive map result of cooperation from those of Figs. 1 and 3

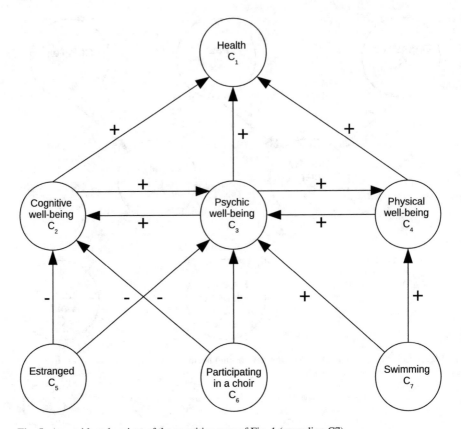

Fig. 5 An accidental variant of the cognitive map of Fig. 1 (regarding C7)

and the influence is positive, the value box is 1; if negative, −1; and if there is no link, 0. Table 1 shows the corresponding matrix for the previous example.

Next, we will use this cognitive map as the basis for representing causal explanations of social relations.

Human beings have been social since their origin and many of our achievements are due to the relationships forged with partners. There are several types of social interaction (Cfr. Gleitman et al. 1981) with some of the most relevant being (a) competition, (b) cooperation and (c) accommodation.

Competition or conflict occurs when two or more individuals or institutions hold opposing views to a problem. This competition has both a positive and a negative aspect: a positive one insofar as it encourages self-improvement and commitment as perhaps only a proposal will succeed; nonetheless, if conflict breaks out in the absence of opportunities, or with an unequal rival, this can cause stress and depression and legitimize social inequalities, showing a negative influence.

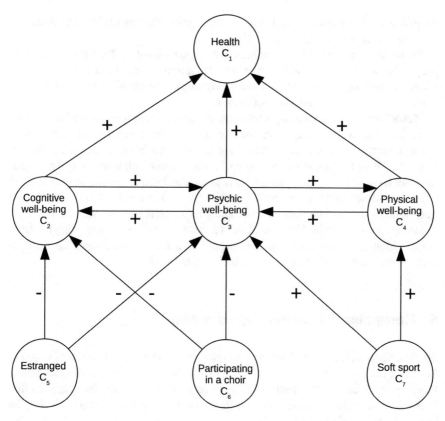

Fig. 6 A consensus map from those of Figs. 1 and 5

The following cognitive map reflects a conflict with that of the Fig. 1, as shown by the matrix scores in bold (Table 2).

This cognitive map could illustrate a physicalist view of health, as opposed to that represented by Fig. 1.

Cooperation occurs when two or more individuals or institutions collaborate to better approach an issue. In cooperative processes, agents often motivate one another to improve a goal. Cooperation is perhaps one of the top forms of socialization, allowing for more and better challenges.

Two cognitive maps approaching a central node from different points of view, so that they can be integrated into a broader and explanatory one, illustrate a case of cooperation. For instance, the following knowledge, illustrated by this cognitive map is compatible with that appearing in Fig. 1. Thus, we can obtain a new and extended cognitive map from the integration of those of Figs. 1 and 3, as shown in Fig. 4.

Finally, the final social interaction we will describe is accommodation. This occurs when there is no cooperation or confrontation; i.e., when a person or institution superficially disagrees with another on how to label a node. That occurs, for instance, if a person loves water sports and modifies the cognitive map of the Fig. 1 by replacing 'running' with 'swimming' in node C7.

In this case, a mediator can suggest a consensus solution for the two different proposals, positing, for instance, that both 'swimming' and 'running' are types of sport, and that only sport practiced in a gentle way is beneficial for the health, giving rise to the agreement graph shown in Fig. 6.

If mediation is unsuccessful, arbitration arises, imposing a well-founded solution that is binding on both parties. For instance, let us suppose that someone labels node C5 as 'Feel marginalized' (in opposition to the previous label, 'Estranged'). We can think that 'Feel marginalized' entails a possible external influence that 'Estranged' does not have and thus, that perhaps there is a radical disagreement on how to label the node with one mark or the other. An arbitrator can force a solution by proposing a new name for the aforementioned node, appealing to a scientific work, an institutional report or any other authorized source. In this case, for instance, suggesting the label 'Ethnic identity' based on a scientific paper that set a causal relationship between said concept and cognitive wellness (Phinley et al. 2001).

4   Resemblance Between Cognitive Maps

As shown above, social relations are detected by locating nodes with links of opposite influence (conflict) or nodes labeled with names that are hyponyms of other terms, unifying them both with a more general word (consensus). This fact suggests that both semantic (meaning) and structural (link) facets are relevant when analyzing the similarity of causal graphs (Fig. 7).

From a semantic perspective, linguistic relations of synonymy, antonymy, hyponymy or hypernymy are important for detecting similarity or distance between the meaning of the names labeling nodes, as they point out, respectively, similarity in meaning, opposite meaning and meaning inclusion. Although linguists sometimes speak about perfect synonymy, in most cases it is a degree relation, as some words are more synonymous than others with a target term. For instance, 'flesh-colored' and 'maroon' are synonyms of 'red', but most people would say that the first word is more synonymous with 'red' than the second one. Similarity measures help to determine to what degree a word bears a resemblance with another. One criterion may be to recursively define a word as the set of words used to define it and to establish that that two words are more similar the more synonyms they share. In Sobrino et al. (2006) we presented a general architecture for implementing an electronic dictionary of synonyms and antonyms capable of assisting in that task.

But a graph is not only characterized by the nodes it has; its position in the global network is also a prevailing factor. A cognitive map differing from another in a central node or in a peripheral node will receive different assessments. Sobrino et al. (2014) suggested a method for calculating a node's degree of centrality, which will also be its degree of relevance. We now go on to briefly describe the underlying notion.

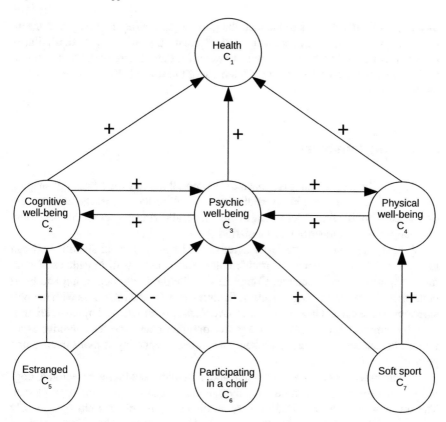

Fig. 7 An arbitration map from those of Fig. 6 and the modification of node C5

A graph $G = (V, E)$ consists of a finite set V of vertices and a finite set $E \subseteq V \times V$ of edges. And edge connects two vertices u and v. The vertices u and v are said to be incident with the edge e and adjacent to each other. Centrality is a function C which maps every vertex v (of a given graph G) to a value $C(v) \in R$. A vertex u is more relevant than another vertex v iff $C(u) > C(v)$. In the aforementioned paper, we approached how to calculate the centrality of a node in terms of different measures of centrality: out degree centrality, closeness, betweenness and eigenvector centrality. Out degree centrality denotes the number of edges pointing out from a node. Closeness is the average length of the shortest path between the target node and all other nodes in the graph: the more central a node is, the closer it is to all other nodes. Betweenness centrality counts the number of geodesic paths (the shortest path between two nodes) passing through a target node. Out Eigenvector centrality measures the popularity of a node in terms of its proximity to other highly connected nodes.

These measures focus on different aspects: for instance, the degree of centrality only takes into account direct connections, but not indirect links. The arithmetic meaning of them all helps to balance the gaps. Aggregating centrality and meaning resemblance measures should help to calculate the similarity or dissimilarity between two cognitive maps and, thus, to establish their degree of conflict or inclusion, contributing to better adjust the intensity of the social behavior represented by them.

5 Closing Summary

In this contribution, we have emphasized two possible roles for citizens in shaping cognitive cities. One function is related to the clothes they wear; the other, to the explanations motivating or justifying their actions, allowing to guess social behaviors like cooperation, competition or mediation.

We have advocated smart textiles as a tool for improving the interaction between humans and their environment, providing data and sensations from which new cognitive experiences may emerge. Companies marketing body monitoring and heart control clothing, luminous garments or flexible displays and interfaces with the outside show some current interactions that have already been achieved by smart textiles. Adaptive smart textiles may well be a source of hitherto unsuspected cognitive experiences unsuspected, enriching citizens' awareness, rendering it more intense and perhaps more invigorating.

Explanations are necessary to understand, and understanding is essential to cognition. Modern cities constitute a highly dynamic scenario where citizens interact. The division of labor, required by very specialized jobs, divides citizens, loosely speaking, into two classes: those who command and those who obey. These roles are possibility assumed without inconvenience if an explanation of what is suggested or imposed is provided.

Explanations are generally offered in long written documents whose understanding requires dedication, interest and a significant amount of time. One Chinese proverb says that 'A picture is worth ten thousand words'. Cognitive maps help to understand a concept through the net of concepts linked to it, pictorially illustrating an abstract idea. A cognitive map recursively defines a concept showing the connections with other concepts influenced by it or being influenced from it. Different cognitive maps reveal different ideas of a concept. If they belong to two different individuals or institutions, they may point to a conflict. Cooperation, conflict or arbitration are social relations that can be diagnosed using cognitive maps.

Similarity or distance between two cognitive maps depends on the meaning of the words labeling the node and on the position of the node in the network. Proximity between words is grasped by semantic relations, such as synonymy, antonymy and hypernymy. Similarity relations provide a calculus for measuring meaning proximity. Centrality measures offer a calculation of the relevance of a node accordingly to its position in the network. Combining them provides us with a tool for ascertaining how one cognitive map is similar to another and, thus, to determine the degree of intensity of a given social behavior.

Cognitive cities constitute a project that is taking shape. In this paper, we have approached two small aspects that can contribute to their development. As Michelangelo said, 'Perfection is not a small thing, but it is made up of small things'. I hope the readers share that opinion.

References

Axelrod R (1976) Structure of decision: the cognitive maps of political elites. Princeton University Press, Princeton

Brandimonte MA et al (2006) Cognition. In: Pawlik P, d'Ydewallwe G (eds) Psychological concepts: an international historical perspective. Psychology Press, Hove

Clark A (1997) Being there: putting brain, body and world together again. MIT Press, Cambridge

Coyle S, Wu Y, Lau K, De Rossi D, Wallace GG, Diamond D (2007) Smart nanotextiles: a review of materials and applications. MRS Bull 32:434–442

Dreyfus H (1992) What computers still can't do: a critique of artificial reason. MIT Press, Cambridge

Dreyfus H, Dreyfus S (1986) Mind over machine: the power of human intuition and expertise in the era of the computer. Free Press, New York

Epstein W, Hatfield G (1994) Gestalt psychology and the philosophy of mind. J Philos Psychol 7(2):163–181

Gleitman H, Gross J, Reisberg D (1981) Psychology, 8th edn. Norton & Company Ltd (2011)

Hauser M, Chomsky N, Fitch T (2002) The language faculty: What is it, who has it, and how did it evolve? Science 298:1569–1579

Hildebrant J, Brauner P, Ziefle M (2015) Smart textiles as intuitive and ubiquitous user interfaces for smart homes. In: Zhou J, Salvendy G (eds) ITAP 2015, Part II, LNCS 9194, pp 423–434

Hoffman G (2012) Embodied cognition for autonomous interactive robots. topics in cognitive science. J Theor Soc Psychol 4(4):1–14

Larkin J, Simon H (1987) Why a diagram is (sometimes) worth ten thousand words. Cogn Sci 11:65–99

Merleau-Ponty M (1945) Phenomenology of perception. Routledge, London (Colin Smith transl. 2002)

O'Flaherty B (2005) City economics. Harvard University Press, Cambridge

Paivio A (1986) Mental representations: a dual coding approach. Oxford University Press, Oxford

Phinley JS, Horenczyk G, Liebkind K, Vedder P (2001) Ethnic identity, immigration and well-being: an interactional perspective. J Soc Issues 57(3):493–510

Sobrino A, Fernández-Lanza S, Graña J (2006) Access to a large dictionary of spanish synonyms: a tool for fuzzy information retrieval. In: Herrera E, Pasi G, Crestani F (eds) Soft-computing in web information retrieval: models and applications. Springer, Berlin, pp 229–316

Sobrino A, Puente C, Olivas JA (2014) Extracting answers from causal mechanisms in a medical document. Neurocomputing 135:53–60

Van Langenhove L, Puers R, Matthys D (2005) Intelligent textiles for protection. In: Scott R (ed) Textiles for protection. Woodhead Publishing, Cambridge

Varela FJ, Thompson E, Rosch E (1992) The embodied mind: cognitive science and human experience. MIT Press, Chicago

Waismann F (1945) Verifiability. Proc Aristot Soc **19**, 119–150 (1945)

Wittgenstein, L (1914–1916) Tractatus logico-philosophicus, Trad. C.K Ogden. Online edition

Alejandro Sobrino is Associate Professor at the Department of Philosophy and Anthropology from the University of Santiago de Compostela (USC), Spain. He holds a Ph.D-degree in Philosophy from the USC and received a MA's degree in Artificial Intelligence from the Polytechnic University or Madrid, Spain. Alejandro Sobrino teaches Philosophy of Language and Philosophy and Artificial Intelligence. His main area of interest is the intersection between language, philosophy and computation. He is secretary of the Journals "Agora. Papeles de Filosofía" and "Archives for the Philosophy and History of Soft Computing".

Possibilities for Linguistic Summaries in Cognitive Cities

Miroslav Hudec

Abstract The shift to smart cities, and further to cognitive cities, should follow citizens' needs, and not only the efficient use of resources. Citizens wish to cooperate in decision-making (or voting) and to be informed about various developments in cities, preferably in comprehensible ways. But, summing up from a large amount of data and gamut of data types is not an easy task. Furthermore, many concepts and predicates are expressed by adjectives and adverbs. Hence, the option are linguistic summaries based on the fuzzy sets and fuzzy logic theory. Other stakeholders in cities (dispatchers, planners, marketers, local government, journalists) may also benefit from this approach. Linguistic summaries are able to verbalize mined information from the data by quantified sentences of natural language such as *most of young citizens have rather negative opinion about topic T* and *most of foreign visits are from countries with medium GDP*. Illustrative examples are focused on informing citizens, managing surveys, explaining development in pollution and traffic, analysing tourist activities. In this way, stakeholders are informed in a concise way about the situation and trends. They can also recognize the effects of regulations which had been brought in. Another benefit is that citizens are better prepared for voting. This contribution also emphasizes the fact that these achievements can be realized without collecting sensitive data from citizens using them as sensors (except volunteers, which prefer simpler data collection). Moreover, exchange of summaries is not as demanding as exchanging sensitive data.

1 Introduction

Each city consists of local authorities, dispatchers, journalists, citizens and a (large) variety of more-or-less available (precise and imprecise) data. The common problems, among others, are optimizing traffic, distribution of water and electrical energy,

---

The original version of this chapter was revised: Affiliation of author "Prof. Miroslav Hudec" has been updated. The correction to this chapter is available at
https://doi.org/10.1007/978-3-030-00317-3_11

M. Hudec (✉)
Faculty of Organizational Sciences, University of Belgrade, Belgrade, Serbia
e-mail: miroslav.hudec@euba.sk; hudec@fon.bg.ac.rs

© Springer Nature Switzerland AG 2019
E. Portmann et al. (eds.), *Designing Cognitive Cities*, Studies in Systems,
Decision and Control 176, https://doi.org/10.1007/978-3-030-00317-3_3

decreasing pollution. The concept of a *smart city* is mainly focused on ensuring such efficiency (Finger and Portmann 2016). Operations research methods have been used to find the best solution for these problems. This way is significant for improving the living conditions in urban systems, but it is simply not sufficient. We should go beyond smart cities and include social aspects like citizen participation, because urban systems are not merely technical artefacts. This direction leads to the concept *cognitive city*.

When we reduce the use of Information and Communication Technologies (ICT) to purely technical tasks such as control and optimization, citizens and other stakeholders like journalists are excluded. It might cause their reluctance to cooperate in the enhancement of living conditions (e.g. in surveys and voting). Furthermore, unpopular decisions have to be taken once in a while. If citizens do not recognize reasons for such decisions, the relationship with citizens will suffer.

But how can we efficiently include and motivate citizens? An option able to comprehensibly interpret relevant (statistical) figures and facts should be considered. Furthermore, we need to communicate with citizens and measure their options and feedback (Perticone and Tabacchi 2016), but they are not always able to express their observations and opinions by numbers, or by one linguistic term from the set of predefined terms.

Dispatchers are overwhelmed with sensor data. These data cover a variety of areas such as traffic, eldercare and energy consumption. Such data can be used as a basis for (immediate) decisions (for example, summing up from eldercare sensors (Wilbik and Keller 2013), or as a valuable overview for service providers and customers (e.g. explaining energy consumption behaviour by short sentences of natural language (van der Heide and Trivino 2009). Further options could be summaries on dashboards for leading workers in departments.

Data summarization is concerned with finding a compact description of the data set. It can be efficiently realized by statistical methods. These methods summarize the essential information from a data set into a few numbers (Campbell and Swinscow 2009). Methods such as means, medians and deviations provide valuable information about the data. Furthermore, means can be expressed by a large family of functions covering the area between conjunction and disjunction (Beliakov et al. 2007). However, this way is less legible for citizens having lower level of statistical literacy. Hence, we should provide summaries which are not as terse as the means (Yager et al. 1990). Graphical interpretation is also a valuable way of summarizing, but cannot always be effective (Lesot et al. 2016). Linguistics is an interesting alternative. In addition, a linguistically summarized sentence can be read out by a text-to-speech synthesis system. It is especially welcome when the visual attention should not be disturbed (Arguelles and Triviño 2013), or in communication with disabled citizens. Apparently, these advantages hold when the resulting data summaries are of high quality (Hudec 2017).

Kaltenrieder et al. (2015) perceived that smart city is based on technology developed to constantly improve interaction with citizens by enhancing cooperation between humans (or more precisely stakeholders) and machines. We believe that suitable interaction can be realized by Linguistic Summaries (LSs), which are based on the *computing with words* concept introduced by Zadeh (1996, 2012).

In this chapter we are focused on possibilities of LSs to help the paradigm shift from smart to cognitive cities. First, aspects and challenges are discussed, followed by an overview of theoretical background of LSs. Subsequently, possible applications of LSs are examined and supported by illustrative examples. Finally, this study concludes with a summary and further perspectives.

2 Aspects and Challenges

If we would like to go beyond the *smart city* concept, we should include sociological aspects and perspectives (Finger and Portmann 2016). Urban systems are not merely technical artefacts; citizens are also a crucial element. It implies that we should consider city as a complex socio-technical system.

Fuzzy logic– one of soft computing methods– may support this goal. Soft computing is often declared as a human-like data and information processing. To be precise, a human brain consists of two hemispheres. The left one is dedicated to logic, mathematics and analysis, whereas the right one is devoted to creativity and summarizing form observations. Analogies have been established between left hemisphere and hard computing, and between right hemisphere and soft computing (Ovaska 2005). Massive connections between hemispheres integrate these two computing paradigms into a highly effective brain.

In tendencies from smart to cognitive cities we recognize the same. Sensors, measurement units, data storages, networks and sharp rules can be viewed as a hemisphere focused on hard computing. The crucial element of a smart city is stakeholders. People communicate, observe, compute and conclude without precise measurements and calculations. They realize these operations by linguistic terms which are vague, but on the other hand effective. We can say that this may be expressed as a hemisphere focused on soft computing. Inspired by the natural connections between hemispheres in a brain, we can establish the interface between these two hemispheres in cognitive cities.

Before we proceed to the main part of the chapter, let us raise several aspects and challenges, which could be supported by linguistically summarized and interpreted data. One of challenges is adapting to the impacts of climate change. We emphasize term adapting, because people prefer high comfort of life, even though it is highly demanding for all kinds of resources. In cold winters people claim about pollution, but are annoyed when local governments try to reduce the number of cars on streets. Possibilities to cope with such problems are motivation for alternative commuting (Huang et al. 2016) or adjusting public transport lines to the real needs (Reinau 2016). Linguistic summaries might be helpful in these directions.

In modern cities elderly people usually live separated from the younger. In the case of health issues, no one is nearby to help them. The impact of climate change (especially during heavy summer days) is visible on parameters such as blood pressure, hearth rate, sweating, indisposition. A solution is gadgets which measure changes in health conditions and interpret them to doctors (Doukas and Maglogiannis 2012).

Ways for linguistically summarizing medical data have been already examined in e.g. (Wilbik and Keller 2013; Wilbik et al. 2011), although not presented as a feature of cognitive city.

Local governments, dispatchers, journalists, citizens are stakeholders with different interests in data and information, but with a common goal: to create and maintain place worth for living. Data appear literally from everywhere: sensors, social networks, surveys, official statistics. One aspect of data is semantic uncertainty. Many data are vague and imprecise by their nature. This imprecision is different from measurement errors, and it is best modelled by the concept of fuzzy numbers (Viertl 2011). We could say that the values of attributes are not always known with the sufficient precision to justify the use of traditional ways for storing and managing them. Conclusions could be different when we neglect fuzziness in the data (Hudec 2015).

The second aspect of semantic uncertainty is related to the concepts in tasks of mining information, classification, querying, summarizing. This uncertainty is not based on randomness or probability; it cannot be expressed by crisp concepts and sets. Vague terms *high*, *cheap*, *around m* ($m \in \mathbb{R}$), heap, most of, few and so forth share three interrelated features of vagueness (Keefe 2000): admit borderline cases, lack sharp boundaries and are susceptible to sorties paradoxes. Vagueness related to the description of the semantic meaning is called fuzziness (Zimmermann 2001).

Most of the interaction between citizens and local governance is usually carried out by natural language (Perticone and Tabacchi 2016). For example, in the eInforming stage of community-building processes unidirectional (top-down) channel informs citizens about relevant policies, projects, Acts, suggestions and news (Terán et al. 2016). Besides these information, citizens should be also informed about relevant figures in easily understandable and interpretable ways.

Generally, we recognize two interrelated applications of LSs: LSs for informing citizens, and LSs expressing relevant knowledge from the collected data for local authorities and executive workers. Before we proceed to the main topic, we recall the basic concepts of LSs.

3 Linguistic Summaries: An Overview

LSs were introduced by Yager (1982). Since then, many research efforts have been conducted, which resulted in numerous theoretical improvements and applications. An overview of recent development in this field can be found in Boran et al. (2016). Prototype forms (protoforms) of linguistic summaries can be divided into the following main groups (Lesot et al. 2016):

- Classic protoforms express summaries related to particular attributes on the whole data set and relational knowledge among attributes by quantifiers such as *most of*, *about half, few*.

- Protoforms of time series express behaviour of attributes over time. These summaries are divided into summaries describing one time series (Kacprzyk et al. 2008) such as *most trends of energy consumption are of medium variability*, and into summaries considering several time series together (Almeida et al. 2013) such as *most of citizens have high participation regarding policy P most of the time*.
- Temporal protoforms explain mode of behaviour. This kind of summaries is illustrated by sentence *regularly energy consumption is high in mornings*. The term *regularly* describes the extent, which a summary holds considering temporal adjustment (Moyse et al. 2013).

In this chapter we are focused on the classic protoforms due to the space limitation and the fact that these protoforms cover basic needs for other prototype forms.

The truth value (validity) of the basic structure of LSs is calculated as Yager (1982)

$$v(Qx(P(x))) = \mu_Q(\frac{1}{n}\sum_{i=1}^{n}\mu_S(x_i)) \tag{1}$$

where n is the cardinality of a considered set (e.g. number of tuples in a database), $y = \frac{1}{n}\sum_{i=1}^{n}\mu_S(x_i)$ is the proportion of tuples that fully or partially satisfy (atomic or compound) summarizer S and μ_Q is the membership function of chosen relative quantifier. Validity gets value form the unit interval.

Relative quantifiers *most of*, *about half* and *few* are plotted in Fig. 1. Quantifier *most of* is expressed by an increasing function, quantifier *about half* is modelled as a convex triangular or trapezoidal fuzzy number and quantifier *few* is explained by a decreasing function. In this way, linguistic meanings of these terms are preserved.

By Eq. (1) we are able to compute, whether summaries such as: *most of citizens cooperate in surveys* and *most of districts have almost zero change in pollution* holds.

Furthermore, we can relax the linguistic meaning of quantifier *most of* with terms *at least few*, *at least about half*. In the opposite direction, we can strength its meaning to very restrictive, but still flexible quantifier: *almost all*. These quantifiers are shown in Fig. 2. We observe that all of them are expressed by increasing functions. Such quantifiers can be used in searching for critical or best entities, e.g. *find buildings where most of {high housing costs, high energy consumption, high pollution, low*

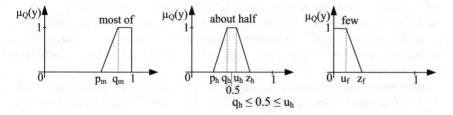

Fig. 1 Relative fuzzy quantifiers *most of*, *about half* and *few*

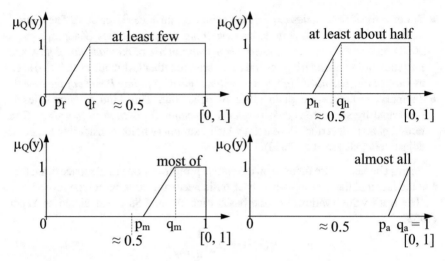

Fig. 2 Relative fuzzy quantifiers *at least few, at least about half, most of, almost all*

suitability of living} and at least few of {elderly households, long distance to centre, medium aged-buildings} are satisfied.

All these quantifiers can be formalized by non-linear functions: Gaussian for quantifier *about half* and sigmoid function for quantifiers *at least few, at least about half* and *most of*. In practice, the linear form is used due to its simplicity. Moreover, for Gaussian or bell shaped functions membership degrees appear on the whole domain, although asymptotically approaching the value 0 when the proportion y (1), (2) is far form the modal value 0.5. Thus, it seems that quantifier *about half* covers the whole domain (unit interval), which does not correlate with the common-sense reasoning.

A more complex type of LS is a summary with restriction. Calculation of its validity (truth value) is realized as Rasmussen and Yager (1997)

$$v(Qx(P(x))) = \mu_Q\left(\frac{\sum_{i=1}^{n} t(\mu_S(x_i), \mu_R(x_i))}{\sum_{i=1}^{n} \mu_R(x_i)}\right) \qquad (2)$$

where variables have the same meaning as in (1), $\mu_R(x)$ is a membership degree to the predicate, which restricts relevant part of a data set and t stands for t-norm.

In this way, we are able to evaluate summaries such as: *most of young citizens have rather negative opinion about suggestion S* and *most of old houses have high gas consumption.*

Restriction and summarizer can by atomic or compound predicates, i.e. consist of several atomic ones merged by logical connectives. In the case of *and* connective, the suitable aggregation is minimum t-norm (Hudec 2016). An example is summary *almost all middle-aged **and** highly educated persons have high participation rate in surveys.*

Regarding the quality of summaries, we should be careful when concluding from LSs with restriction. These summaries might be trapped into outliers due to possible very low coverage of tuples in R and S, even though the validity is high. For instance, let only 2 of 5.10^6 tuples fully meet R and the same tuples fully meet S, then the validity (2) gets value of 1 leading us to the conclusion based on outliers. In the basic structure (1) cardinality of a data set is in denominator, i.e. low coverage is reflected in low validity.

Evidently, this problem of the LSs should not be neglected. According to Hirota and Pedrycz (1999), the following features are essential for measuring the quality of data summaries: validity, generality, usefulness, simplicity and novelty. Wu et al. (2010) have proposed equations for calculating these measures for linguistic summaries in order to convert them into the fuzzy rules. A simplified quality measure aggregating validity and coverage is suggested in Hudec (2017) as

$$Q_c = \begin{cases} t(v, C), & C \geq 0.5 \\ 0, & \text{otherwise} \end{cases} \quad (3)$$

where t stands for t-norm and C for coverage. Concerning the suitable t-norm, experiments have shown that the minimum t-norm is not a suitable one, while product t-norm is preferred. The coverage C explains, how many records influence the validity of LS (2). Calculating coverage for different cases can be found in Hudec (2017), Wu et al. (2010).

The relevant data for summaries may be of different types: precise numbers, (weighted) categorical data, short text or sentence fragments, fuzzy or imprecise data. Further, predicates in summarizer and restriction can be crisp or fuzzy. The meaning and structure of quantified summaries (1) and (2) remains unchanged regardless data types, because in these summaries we operate only with matching degrees to the respective atomic predicates.

The simplest case is sharp predicate for precise data, i.e. *consumption of electricity per month > 500 kWh*. Matching degree is either 0 or 1.

For the fuzzy condition on crisp data we assign matching degree by membership functions. For example, distance of 10.2 km belongs to the concept *short distance* (*SD*) expressed as

$$\mu_{SD}(x) = \begin{cases} 1 & x \leq 10 \\ \frac{12-x}{12-10}, & 10 < x < 12 \\ 0, & x \geq 12 \end{cases}$$

with degree 0.9.

The other possibilities are illustrated in the subsequent sections.

Fuzzy concepts can be expressed by non-linear functions. The main problem is assigning suitable shape. The slope of linear function is the same at each point, whereas in non-linear function slope varies in different points. In the aforementioned example, user declares that distance up to 10 km is without any doubt short, distance greater than 12 km is for sure not short and distance in interval (10, 12) is partially short. Regarding the shape, user usually cannot say which non-linear function and parameters suitably explain fuzzy concept. Hence, linearity is a desirable simplification.

The meaning of fuzzy relative quantifiers is generally accepted and therefore parameters (for linear or non-linear functions) can be defined in advance for variety of tasks. On the other side, fuzzy concepts are sensitive to the contexts, where they are applied and therefore users (in our case stakeholders in cities) would welcome simplified functions defined by easily understandable parameters.

3.1 Matching Degrees for Fuzzy Data

For calculating matching degrees to fuzzy predicates possibility and necessity measures are options. Comparator $\leq$ is fuzzified by *fuzzy possibly less than* and *fuzzy necessarily less than* comparators (Galindo et al. 2006). Analogously holds for other comparators. Fuzzy data are generally divided into two groups: continuous, which are expressed by fuzzy numbers; and discrete or countable, which cover, for instance, weighted categorical data.

Continuous fuzzy data (fuzzy numbers)

In this part, we recall *fuzzy possibly greater than* and *fuzzy necessarily greater than* comparators. Other comparators were straightforwardly extended from the possibility and necessity measures.

The *fuzzy possibly greater than* comparator is calculated as Galindo et al. (2006)

$$\mu_{PFG}(r) = \begin{cases} 1 & m_A \geq m_F \\ \dfrac{b_A - a_F}{(m_F - a_F) - (m_A - b_A)} & m_A < m_F \wedge \\ & \wedge b_A > a_F \\ 0 & b_A \leq a_F \end{cases} \tag{4}$$

where a_A, m_A, b_A stand for the parameters of triangular fuzzy number A and m_F and b_F explain fuzzy concept F shown in Fig. 3.

Analogously, the *fuzzy necessarily greater than* comparator is calculated as Galindo et al. (2006)

$$\mu_{NFG}(r) = \begin{cases} 1 & a_A \geq m_F \\ \dfrac{m_A - a_F}{(m_F - a_F) - (a_A - m_A)} & a_A < m_F \wedge \\ & \wedge b_A > a_F \\ 0 & m_A \leq a_F \end{cases} \tag{5}$$

where parameters have the same meaning as in (4) and in Fig. 3.

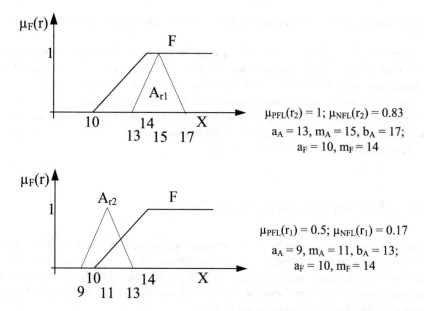

$\mu_{PFL}(r_2) = 1$; $\mu_{NFL}(r_2) = 0.83$
$a_A = 13$, $m_A = 15$, $b_A = 17$;
$a_F = 10$, $m_F = 14$

$\mu_{PFL}(r_1) = 0.5$; $\mu_{NFL}(r_1) = 0.17$
$a_A = 9$, $m_A = 11$, $b_A = 13$;
$a_F = 10$, $m_F = 14$

Fig. 3 *Fuzzy possibly greater than* and *fuzzy necessarily greater than* comparators for two fuzzy numbers A_{r1} and A_{r2}

The possibility measure is considered as an optimistic one, whereas the necessity as a pessimistic one, i.e. $\mu_{NFL}(r) \leq \mu_{PFL}(r)$. This is also illustrated in Fig. 3 for two fuzzy numbers A_{r1} and A_{r2}.

Discrete fuzzy data

Possibility and necessity measures can be applied on fuzzy data expressed by discrete values. The former measure is adjusted to the countable set for the comparator *fuzzy possibly equal to* in the following way (Hudec 2015)

$$Poss(\mu_A(r), \mu_C(r)) = \max_{i \in N_T}[t(\mu_{A_i}(r), \mu_{C_i}(r))] \qquad (6)$$

where index A stands for fuzzy data (opinion of respondent r), C for fuzzy concept, i for ith linguistic term, N_T for the set of linguistic terms and t for t-norm (usually minimum function is applied).

Analogously, comparator *fuzzy necessarily equal to* can be expressed as

$$Nec(\mu_A(r), \mu_C(r)) = \min_{i \in N_T}[\max(1 - \mu_{C_i}(r)), \mu_{A_i}(r)] \qquad (7)$$

Let us express predicate *positive opinion* as fuzzy set $PO = \{(\text{high}, 0.75), (\text{very high}, 1)\}$. If citizen (respondent r) answered in the following way: $A_r = \{(\text{high}, 0.4),$

(very high, 0.6)}, then the matching degree to *possibly equal to positive opinion* is 0.6.

Due to $\sum_{i=1}^{n} \mu_{A_i}(r) = 1$ these fuzzy data are not normalized fuzzy sets, except singleton answers. This option is not the ideal one, because we may expect that answer {(medium, 0.25), (high, 0.75)} is worse than {(high, 0.75), (very high, 0.25)}, but solutions are the same. This problem can be solved by normalized fuzzy sets, i.e. $\max_{i=1,...,n} \mu_{A_i}(r) = 1$, similarly as it works for fuzzy numbers. It may cause better adjustment to these answers, but it influences the semantic meaning. Evidently, further examination in this field is advisable.

3.2 Matching Degrees for Short Texts

Attributes can be expressed by short sentences or sentence fragments. For instance, an employer explains productivity of respective workers in a local administration as: more average than small, average, rather high, high, etc. When we wish to find out summarized overview about the productivity, we need to find similarity between term(s) in the predicate and recorded term(s).

A possible form of similarity between terms is the membership function (Niewiadomski 2002)

$$\mu_{RW}(w_1, w_2) = \frac{2}{N^2 + N} \sum_{i=1}^{N(w_1)} \sum_{j=1}^{N(w_1)-i-j} h(i, j) \tag{8}$$

where $N(w_1)$ and $N(w_2)$ are numbers of letters in words w_1 and w_2, respectively, $N = \max\{N(w_1), N(w_2)\}$, binary function $h(i, j) = 1$ if a subsequence containing i letters of word w_1 and beginning at the jth position in w_1 appears at least once in word w_2, otherwise this function equals to 0.

When attribute contains short sentences or fragments, we should compare them with term(s) in summarizer and restriction. It can be realized in the following way (Niewiadomski 2002)

$$\mu_{RZ}(z_1, z_2) = \frac{1}{N} \sum_{i=1}^{N(z_1)} \max_{j=1,...,N(z_2)} \mu_{RW}(w_i, w_j) \tag{9}$$

where $N(z_1)$ and $N(z_2)$ are numbers of words in sentences z_1 and z_2, respectively, $N = \max\{N(z_1), N(z_2)\}$, w_1 and w_2 have the same meaning as in (8).

We expect that $\mu_{RW}(w_1, w_2)$ reflects the fact that the more common the subsequence of given words are, the more similar they are, but the main drawback lies in not reflected semantic similarity (Duraj et al. 2016). Terms *small* and *low* in some contexts are synonyms, but this approach cannot cover it. Possible solutions are construction of similarity matrix between all expectable terms and partial order or lattice among linguistic terms.

3.3 Constructing Membership Functions

Construction of fuzzy sets is a sensitive task, because chosen parameters of fuzzy sets influence the solution. Thus, membership functions should be carefully constructed. Fuzzy sets for concepts such as *short distance* and *high population density* depend on the context and geographical areas, where they are employed.

The same holds for interpreting numeric data. For instance, temperature in a weather forecast can be explained linguistically: it is convenient to take short-sleeved shirts. It is a valuable information for local population. When temperature is explained by numbers, e.g. 25 °C, then tourists from the warmer countries know that taking a jacket is reasonable. Linguistic interpretation is not always the best option to convey information to all people.

LSs are a suitable way for explaining summaries, retrieving data and mining from the data when flexible concepts in R and S are properly formalized by membership functions to cover particularities of the considered city and region. We cannot assign the same parameters to concept *low population density* to cities in China and Switzerland, for example. This observation holds for any task dealing with linguistic terms and membership functions. For instance, linguistic terms in query condition *short distance and low price* for finding suitable hotel have the same shape (both concepts are expressed by a decreasing function), but different parameters for student and businessman. Price 32 € might be partially acceptable for student, whereas this price is without any doubt low for businessman.

Generally, there are two main aspects for constructing fuzzy sets. In the first aspect, users choose shape and parameters for each linguistic term according to their best knowledge. The second aspect deals with the mining these parameters from the data. We usually know the domain of considered attribute (all theoretically possible values), but the collected data may cover only a part of domain. Thus, membership functions' parameters should reflect this fact (Hudec et al. 2014). For instance, number of days when the measured pollutant exceeds its limit may be any value from the [0, 365] interval of integers. Therefore a database must allow storing any value from this domain. In practice, measured values cover only a part of this domain. Moreover, these parts differ among counties and cities. Thus, *medium pollution* is not the same for different cities. Furthermore, creating uniformly distributed granules *small*, *medium* and *high* considering the smallest and the highest collected value is not adequate for the highly unbalanced data distribution. In such cases we can apply other ways: the statistical mean based method (Tudorie 2008) or the logarithmic transformation of domains and evaluate quality measures.

Membership functions can be constructed by merging both aspects. In the first step, parameters of fuzzy sets are calculated from the current content of a database.

In the second step, users can modify these parameters if they are not satisfied with the suggested ones. This process is more tedious, but on the other hand, users see how their expectations correlate with the data.

Other branches of soft computing can help in assigning parameters to membership functions. When the structure of relevant summaries and historical data are at disposal, then parameters of fuzzy sets can be calculated by neural networks or adaptive neuro-fuzzy inference systems. Due to the limited space, applicability of other soft computing branches is not further examined.

It is worth noting that if we want to develop an easy to use approach for broad audience and less demanding for the computation (e.g. on websites), then sophisticated approaches supported by neural networks or evolutionary algorithms are not the best options. Computing parameters of fuzzy sets and validities of LSs may be, in these cases, realized through an additional functionality of the existing SQL-like query engines.

3.4 Mining and Interpreting Summaries

Generally, two options exist: calculating validity of a particular summary of interest or revealing all summaries with high validity (and quality). The latter has the following structure (Liu 2011)

$$
\begin{aligned}
&\text{find } Q, S, R \\
&\text{subject to} \\
&Q \in \overline{Q} \\
&R \in \overline{R} \\
&S \in \overline{S} \\
&v(Q, S, R) \geq \beta
\end{aligned}
\tag{10}
$$

where $\overline{Q}$ is a set of quantifiers of interest, $\overline{R}$ and $\overline{S}$ are sets of relevant linguistic terms for restriction and summarizer, respectively, and β is a threshold value from the (0, 1] interval. Each feasible solution produces a linguistic summary ($Q^* R^*$ are S^*). Moreover, mining summaries by this equation can be extended with an additional constraint: quality (3) should be above the threshold value (Hudec 2017).

Before proceed to the applicability, a short illustrative example demonstrates benefits of LSs.

Example 1 Let the data set contains the following seven citizens-commuters with their respective ages $\{C_1:26, C_2:28, C_3:32, C_4:40, C_5:54, C_6:56, C_7:57\}$. Statistical summarization reveals that the average age (arithmetic mean) is $41.857 \approx 42$, median is 40 and standard deviation is 13.668. If standard deviation is overlooked or person is not skilled with statistical figures, he/she concludes that the representative commuter is middle-aged.

LSs calculated by (10) interpret solution differently. The terms set expressing quantifiers is $\overline{Q} = \{$few, about half, most of$\}$. These terms are uniformly constructed

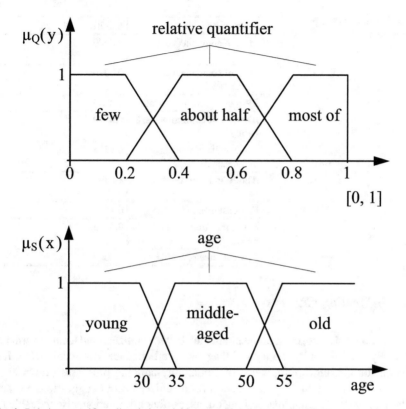

Fig. 4 Relative quantifiers, linguistic variable *age*, and their respective labels

on the unit interval as is plotted in Fig. 4. The summarizer for linguistic variable *age* consists of terms $\overline{S} = \{young, middle\text{-}aged, old\}$. In this example, we use basic structure of LS (1), i.e. the second constraint in (10) does not appear. The fuzzy granules of variable *age* are shown in the same figure. Possible LSs and their validities are shown in Table 1. When the threshold value $\beta = 0.75$ is applied, only LSs marked as bold are shown to user.

At first glance it is obvious that middle-aged commuters represents minority. This way of interpreting results is understandable for broad spectrum of stakeholders from top managers in a local government to general public. □

4 Possible Applicability in Smart Cities

This section is dedicated to possible applicability in areas, where this approach may be helpful.

Table 1 Commuters age expressed by LSs

Linguistic summary	Validity
Most of commuters are young	0.0000
Most of commuters are middle aged	0.0000
Most of commuters are old	0.0000
About half commuters are young	**0.8570**
About half commuters are middle aged	0.1425
About half commuters are old	**1.0000**
Few customers are young	0.1430
Few commuters are middle aged	**0.8575**
Few commuters are old	0.0000

4.1 Informing Citizens

As was indicated afore, most of the interaction between citizens and local governance is usually carried out by the natural language (Perticone and Tabacchi 2016). It is relevant for, e.g. eInforming stage of community-building processes (Terán et al. 2016). In this stage, local government informs citizens about suggestions for new Acts and accepted regulations, among others. Further information, also valuable for citizens, are statistical figures and opinions (managed by questionnaires or received from statistical offices). Data summarization from numbers by statistical methods is a convenient way, but rather understandable for citizens with higher statistical literacy. On the other hand, summaries by short quantified sentences of natural language handle different data types and are understandable for majority of citizens. Such summaries might help citizens in discussing, voting and also in recognizing effects of regulations.

Informing citizens regarding pollution, comparing their cycling records and managing surveys are discussed in respective subsections.

4.2 Surveys Among Citizens

Surveys are deemed to be a reliable barometer of citizens' opinions about developments in various aspects of society. Surveys on the national level collect data form the whole population or representative sample, whereas surveys in cities realize the same, but on the local level. Thus, we can use knowledge already achieved in official statistics. Reluctance to cooperate in surveys is a frequent issue due to feeling of response burden and lack of recognized benefits (Giesen 2011). Possible solution

Table 2 Time required for filling questionnaire Q_q

Respondent	Time (min)
Respondent 1	36
Respondent 2	22
Respondent 3	41
Respondent 4	38
Respondent 5	50
Respondent 6	39
Respondent 7	44
Respondent 8	10
Respondent 9	37
Respondent 10	48

is motivation by tailored survey designs and by interpreting figures from surveys in easily understandable ways (Hudec and Torres Van Grinsven 2013; Torres van Grinsven 2015). With such support, citizens may recognize the point of surveys.

The next example illustrates benefits of summaries by short quantified sentences in recognizing the most demeaning questionnaires (Hudec 2016).

Example 2 An agency launching surveys by questionnaires wishes to know, which of them are the most demanding for respondents. Let us for the illustrative purpose have several questionnaires Q_i answered by respondents. Questionnaires may be of different complexity (number and structure of queries). In order to consider this fact, fuzzy set *high response time* is constructed for each questionnaire considering recorded response times. Response times for Q_q are shown in Table 2. The fuzzy sets *high response time* for this questionnaire is expressed as

$$\mu_{HR_q}(x) = \begin{cases} 0 & x \leq 35 \\ \frac{x-35}{40-35}, & 35 < x < 40 \\ 1, & x \geq 40 \end{cases}$$

applying the fuzzy partition as in Ruspini (1969) by the uniformly covering domain (for the simplicity data distribution is not considered). The next option for constructing fuzzy set *high response time* is asking designers of questionnaires to express their tacit knowledge regarding expected high response time for respective questionnaires.

The query for recognizing demanding questionnaires is *select questionnaire where most of respondents have high response time*. Using fuzzy set HR_q, basic structure of LSs (1) and quantifier *most of* with parameters $p_m = 0.5$ and $q_m = 0.85$ (Fig. 1) we calculated validity for Q_q as 0.286. Let us assume that the validity for Q_m is 0.895. These validities explain that Q_m is significantly more demanding and should be re-designed, whereas Q_q is not the priority. Otherwise, respondents may not cooperate in the future surveys or put less attention to questions, which might cause lower quality of surveyed data. □

The answer emphasises which questionnaires are more demanding and therefore should be re-designed. Designers receive summarized information about questionnaires. Sensitive data regarding respondents and their respective answers remain undisclosed. This information can be exchanged among different departments, agencies focused on survey design and even cities to share experiences. Furthermore, this way illustrates another benefit of linguistic summaries: quantified query condition for selecting relevant entities.

Another problem in surveys is the requirement for precise answers. Usually, respondents should provide the requested answers by exact numbers or choose one term from a set of predefined linguistic terms (categories) in a Likert scale (Likert 1932) way. It is a suitable option for the fast data processing. In surveys citizens also welcome space for expressing observations by sentence fragments, weighted categories and imprecise numbers. For instance, *a bit smaller*; *negative with 0.75 and neutral with 0.25* (to explain opinion *more negative than neutral* by fuzzified Likert scale (Li 2013); *around 120 but for sure not less than 95 and not greater than 150* (by fuzzy number (Hudec 2016), respectively. Regarding the first, each relational database is able to store such data. Concerning, the second and the third, relational databases cannot store these data due to the first normal form Date (2006). The solution can be the fuzzy extension of relational databases called fuzzy meta model (Hudec 2016; Škrbić et al. 2011). In this way each classical relational database is able to manage fuzzy data. Fuzzy number *around 120 but for sure not less than 95 and greater than 150* is expressed as triangular fuzzy number $A_1(95, 120, 150)$. Clear opinion, e.g. 110 is converted to $A_2(110, 110, 110)$ for handling arithmetic operations among fuzzy numbers. These operations are discussed in, e.g. (Bojadziev and Bojadziev 2007).

The next example explains mining summaries from the imprecise citizens responses regarding the willingness to support new activity in a city. More precisely local authorities wish to see, whether high opinion influences willingness to support community activities with higher amounts.

Example 3 Local authorities launched a survey regarding citizens opinion about activity G and willingness to financially participate in a linked up community project. The hypothesis is that *most of citizens with positive opinion would provide higher amounts*. The validity of this hypothesis can be checked by LSs with restriction (2). Collected data could be stored in a transactional relational database extended with the fuzzy meta model. Because the focus in on mining summaries, aspects related to managing fuzzy data are skipped.

For the sake of simplicity collected data are shown in Table 3, columns: respondent, age, opinion and amount. For example, parameters (70; 80; 95) mean that respondent is willing to provide amount of 80, but not less than 70 and not more than 95. This opinion is expressed as a triangular fuzzy number. We see that $r\,2$ does not have any doubt regarding amount he/she is willing to pay. Response "M:0.6; H:0.4" for the *opinion* attribute means that the respondent's inclination is a bit more towards *medium* than *high*. This value is not a normalized fuzzy discrete set. Issues regarding the normalization are mentioned in Sect. 3.1.

The concept *positive opinion* is expressed as fuzzy set $PO = \{(\text{high}, 0.75), (\text{very high}, 1)\}$ and the concept high amount (HA) is expressed as

Table 3 Respondents answers in a survey and respective matching degrees

Respondent	Age	Opinion[a]	$\mu_{HO}(r)$	Amount	$\mu_{PossHA}(r)$	$\mu_{NecHA}(r)$
r 1	28	M:0.6; H:0.4	0.4	70; 80; 95	0.27	0
r 2	32	M:0.5; H:0.5	0.5	100; 100; 100	0.5	0.5
r 3	65	H:0.4; VH:0.6	0.6	90; 100; 110	0.6	0.4
r 4	45	VH:1	1	110; 120; 130	1	0.8
r 5	58	M:0.5; H:0.5	0.5	100; 110; 120	0.8	0.6
r 6	41	L:0.8; VL:0.2	0	55; 60; 75	0	0

[a]VH stands for very high opinion, H for high, M for medium, L for low, VL for very low

$$\mu_{HA}(x) = \begin{cases} 0 & x \leq 80 \\ \frac{x-80}{120-80}, & 80 < x < 120 \\ 1, & x \geq 120 \end{cases}$$

In this example, two operators for calculating matching degrees are used: possibility and necessity. For the former, the LS is *most of possibly positive opinions would possibly provide high amounts*. Matching degrees to possibly greater than HA (4) are shown in Table 3, column $\mu_{PossHA}(r)$. The validity of this summary is 1. The next step is calculating quality. Coverage C in quality measure (3) is equal to 1, thus $Q_c = 1$. We can say that this summary has proven our hypothesis. This is far form the rigid statistical way, but a suitable option for the fast evaluations. The necessity measure shows lower, but still significant validity (0.52) and high quality $Q_c = 1$.

When local authorities wish to know further relational knowledge such as relation between different age groups and opinions, or age groups and acceptable amounts, LSs can be straightforwardly adjusted. □

This raises the question of the best methods to ask citizens to produce their fuzzy opinions. The simplest way is an interface, which allows selecting one of offered linguistic terms. Number of terms varies according to the considered question. In the real-life tasks, usually three to seven terms cover the domains, with nine being an upper limit according the human limited capacity (Miller 1956).

One can find this method in airports, train stations, banks and so forth. Usually it is a panel with three buttons: smile, neutral face and sad. This method is suitable for places, where people are frequently in a hurry. People simply push one button and proceed somewhere else. Temporal prototype form of LSs is also suitable for interpreting behaviour, i.e. *regularly on Saturdays afternoon satisfaction is low*.

For surveys where citizens can answer whenever they want (in a defined time frame) we can offer more sophisticated methods. In many occasions respondents wish to provide answers like: *more neutral than negative*. This vague answer covers cases from, let say, {(0.1, negative), (0.9, neutral)} to {(0.45, negative), (0.55, neutral)}. We can hardly manage such vagueness and use it in data analyses.

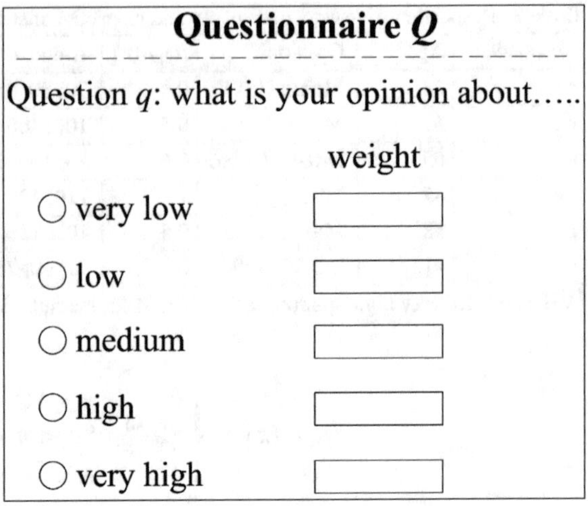

Fig. 5 Collecting weighted categorical answers by input boxes

The interface illustrated in Fig. 5 is more tedious. When respondents select two options, they should also provide numerical weights. On the other hand, these answers better explain vagueness and are easily processed in fuzzy relational databases and in data mining tasks (Hudec 2016).

The better option is sliders which cover the whole domain as is plotted in Fig. 6a. In Fig. 6b and c two possible positions of slider are shown (values 0,...,4 are shown for the calculation purpose). This way is convenient for respondents, because they simply move slider to the desired position. The procedure for converting values from sliders is explained in the following code

```
Select Case lp \position of slider
Case 0
    -> VL:1  \assinging weight 1 to term very low
Case 1
    -> L:1   \assinging weight to term low
...
Case 4
    -> VH:1
...
Case >2 and <3
    lpn=lp-2     \value of slider into [0, 1]
    W_M=1-(lpn)  \calculating weight for term medium
    W_H=lpn      \calculatin weight for term high
    M:W_M        \assigning weight to term medium
    H:W_H        \assigning weight to term high
...
```

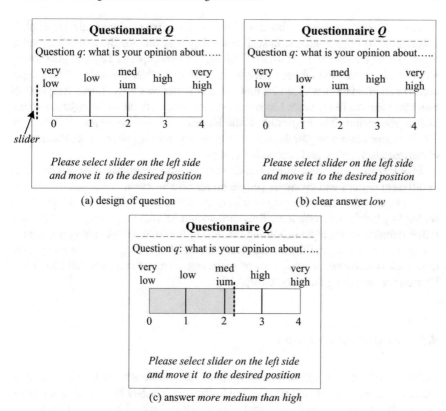

(a) design of question   (b) clear answer *low*

(c) answer *more medium than high*

Fig. 6 Collecting weighted categorical answers by sliders

In a referendum citizens have to mark one answer: Yes or No. It is a bivariate decision making, which is not suitable for all questions. Let 51% of citizens prefer very lightly answer Yes, whereas 49% are clearly against. When all citizens should give their respective answers, then the result of referendum does not correlate with the citizens preferences. The solution is flexible voting (Ladner and Meier 2014) which can be realized by aforementioned sliders. Adopting flexible voting in a cognitive city level is not so demanding as for the country level, where consensus among political parties, Constitutional court and so forth should be achieved. Calculating weights is simpler than in Fig. 6, because we do not have intermediate categories. The answer is position of slider inside one interval. When $l_p/l = 0.5$ (where lp is the slider position and l is the length of interval), it means that citizen does not know which answer to prefer. High appearance of similar answers indicate that something is wrong with the question. The final score is calculated by dividing sum of individual values from the unit interval with number of participating citizens, where aggregated value higher than 0.5 means answer Yes.

The next speculation is related to the fact that citizens cannot always explain observation by crisp numbers. In such cases they may provide the most likely value together with the left and right limits. Such answers are expressed by triangular fuzzy numbers. It means that citizens have to provide more than one value. Three input boxes in an interface is not the best method. The better way is by sliders. Respondent provides the most likely value and moves sliders to the left and to the right to explain his/her uncertainty. The inspiration is the ReqFlex fuzzy query engine (Smits et al. 2013), where fuzzy sets parameters are adjusted by moving sliders over the domain of attribute to set ideal and acceptable values.

In order to adapt the best option real-life experiments are highly advisable, because in different regions and countries people prefer diverse ways.

The next option for collecting opinions are social networks where people more willingly provide their opinions. But, care should be taken regarding the representative sample. An example of mining citizens opinions about launching population census can be found in (Torres van Grinsven and Snijkers 2015). The recognized drawback is interpreting intensities of sentiments. Thus, fuzzy sets and LSs might be used for managing this uncertainty.

4.3 Mitigating Pollution

Pollution is a common sign in many cities, including those without heavy industry in the near vicinity. Significant sources are traffic and obsolete heating systems (low-quality filters, burning waste to avoid paying waste disposal fee, etc.). Concerning the latter, improvements can be realized by education and financial support for more effective heating. Regarding the former, optimization of public transport and supporting bicycling are options. Anyway, people prefer to be informed about pollution and its development in time, preferably by verbal explanations.

In many cities pedestrian and cycling culture (a common way for commuting in the past) is growing. The important aim is encouraging people in this direction, which can be achieved by education, motivation and competition.

4.3.1 Summaries from Sensors

Let us for illustrative example have network of sensors in the whole city, which is divided into districts or precincts. The collected values form sensors can be transformed into the snowflake structure of a data warehouse shown in Fig. 7. It is a dimensional data structure, where each measured value gets its meaning in the intersection of dimensions (Kimball and Ross 2013). This structure is suitable for asking variety of business intelligence questions (Jensen et al. 2010). Our focus is on questions and answers expressed by quantified sentences of natural language.

Data from this structure are shown in Table 4, where measured value can be the maximal daily value, or the other suitably aggregated value.

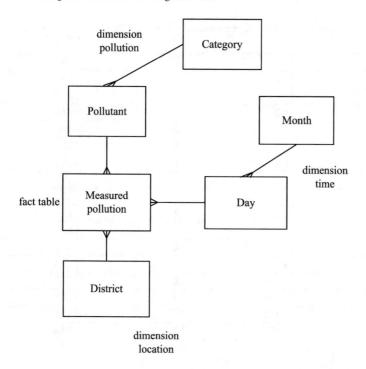

Fig. 7 Dimensional data structure known as *snowflake* scheme for managing pollution

Example 4 An environmental department wishes to recognize districts where most of days measured pollution is high. To solve this task we apply the basic structure of LSs (1). In Table 4 district $d\,1$ contains eight sensors, hence $n_{d1} = 8$. Fuzzy set *high pollution* is expressed as

$$
\mu_{HP}(x) = \begin{cases} 0 & x \le 15 \\ \frac{x-15}{5}, & 15 < x < 20 \\ 1, & x \ge 20 \end{cases}
$$

The validity of this summary for district $d\,1$ is 0.61, whereas for $d\,2$ is 0 when we use parameters $p_m = 0.5$ and $q_m = 0.85$ for quantifier *most of*. Pollution in $d\,1$ is significant, but not severe ($v < 1$). By this LS we can rank districts downwards from the most polluted one. Geographically, district is a polygon. Thus, we can interpret validities on thematic maps. Districts which are severely polluted can be marked with some colour, let us say red, low-polluted districts can be marked with another colour (quite different than the one for expressing severe pollution) and districts which partially meet the summary condition are marked again with some other colour having a gradient from faint to deep hue according to their matching degrees.

When we apply sharp condition: *select districts where pollutant P exceed value of 20*, then both districts are equally considered. Measured pollution of 19.5 μg is

Table 4 Pollution from a snowflake structure shown in Fig. 7 for districts, sensors and days

District	Sensor	Day	Value (μg)
d 1	s 1	1	19
d 1	s 1	2	18.5
d 1	s 1	3	14.2
d 1	s 2	1	17.3
d 1	s 2	2	19.8
d 1	s 2	3	21
d 1	s 3	1	19.5
d 1	s 3	2	19.5
d 2	s 100	1	12
d 2	s 100	2	11
d 2	s 100	3	8.5
d 2	s 100	4	18.2
d 2	s 101	1	21.1
d 2	s 101	2	17.3
d 2	s 101	3	12.8
d 2	s 102	1	11.6
d 2	s 102	2	12.8

not considered as high. It is treated in the same way as pollution of 8.5 μg, because classical sets are not able to distinguish intensity of belonging. □

Someone might argue that we should also consider other factors like population and dwellings density, in other words, to work with the relative high pollution. From the technical point of view this is reasonable, but for citizens' lungs this is irrelevant. Actually, we can adjust parameters of fuzzy set *high pollution* to specific or unexpected wind and air pressure conditions.

In a high-density network of sensors it might happen that several of them are out of order or improperly calibrated. The fuzzy logic approach is less sensitive to these cases, because we work with non-sharp concepts. Assume that in *d 1* sensor *s 2* has not worked third day and the sharp concept is pollution > 20. It means that only in *d 2* pollution above the limit is recorded. Data users may conclude that this district is more polluted. When we launch competition among districts regarding decreasing pollution, or trace development in time, borderline cases should be adequately considered.

The next possibility is linguistically interpreting changes over time. Theory of time series is a mature science field capable to cope with broad variety of tasks and trends. Our intent is to demonstrate possibilities of classic protoforms of LSs in explaining time series.

Table 5 Difference in percentage for pollution between months T_i and T_r (ΔT)

District	Sensor	ΔT (%)
d 1	s 1	−1.9
d 1	s 2	−3.2
d 1	s 3	−10.8
d 1	s 4	2.3
d 1	s 5	−12.1
d 1	s 6	−5.8
d 1	s 7	−7.6
d 1	s 8	−8.1
d 2	s 101	0.2
d 2	s 102	1.1
d 2	s 103	0.3
d 2	s 104	−0.2
d 2	s 105	1
d 2	s 106	0.6

Example 5 Let us have pollution aggregated to the month level for each sensor. Percentage changes of pollutant P between initial and reference month are shown in Table 5.

Usually, in technical systems changes in the left half of the interval are labelled as negative, e.g. *negative small, negative medium* and *negative high*. Correspondingly are named changes in the right part of the interval (Zimmermann 2001). Changes near value of 0 are labelled as *around zero*. In our case, linguistic labels of variable *change in pollution over time* are plotted in Fig. 8.

In order to solve this task we apply basic structure of LSs (1) for each district and (10) for revealing relevant summaries. Hence, quality of summaries is implicitly calculated. The set of possible terms for summarizer is $\overline{S}$ = {high decrease, medium decrease, small decrease, about zero, small increase, medium increase, high increase} on attribute ΔT, which might take any real value.

Applying the same fuzzy set for quantifier *most of* as in Example 4 the following summaries are mined:

- most of sensors in district *d 1* measured small decrease (validity 0.911)
- most of sensors in district *d 2* measured about zero change (validity 0.72)

Improvements are recognized in district *d 1*. It implies that for instance, regulations taken by local government (if any) work. The change in *d 2* is around zero, but keeping in mind that pollution is low, we can say that this district keeps high quality of environment. □

Someone might argue that, for example, January in successive years significantly differ in number of very cold days and predominant winds, and therefore higher

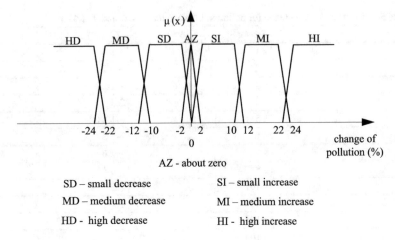

Fig. 8 Fuzzification of the linguistic variable *change*

pollution is acceptable. Theoretically, fuzzy sets can be constructed on the domain of relative pollution (pollution, which consider these facts), but care should be taken, because citizens' lungs definitely do not share this opinion. They will not suffer less if we increase tolerance level.

Unpopular decisions have to be taken once in a while. In some cities reducing comfort of travelling form the home door to the office door by cars is considered as unpopular. But when positive effect (reduced pollution and respiratory problems of small children) is measured and suitably interpreted, then authorities have strong argument. When the mined information from the data show improvements, then citizens can identify with the regulations. Moreover, authorities might offer access to such data warehouses and offer tools for linguistically summarizing data to check, whether claims of authorities corresponds with the data. In addition, such summaries are valuable for journalists and researchers. Another benefit is that protecting data privacy is not required. Sensors measure pollution, which is result of citizens behaviour and industry, not the sensitive behaviour of citizens.

If pollution decreases or is on the stable low level, it seems that regulations work and citizens will further cooperate in improving living conditions. By LSs local authorities and citizens can check situation any time. A dashboard presenting summaries from the previous day or development in time by summaries and graphs could be a valuable first glance about pollution for all stakeholders.

4.3.2 Summaries of Travel Behaviour

The OECD study (OECD 2012) expects the number of vehicles world wide in 2050 will double in comparison with 2012. A modern car produces less pollution compared to car produced 40 years ago. On the other hand, number of modern cars is significantly higher. Many cars will be electric, but electricity must be somewhere produced. The higher number of cars on streets would require extending existing and building new roads. Often, people prefer comfort than protecting nature, otherwise this increase would not be estimated. Initially, the role of cars was to transfer people in the fastest way from the point *A* to the point *B*. Many people consider commuting by cars as the lest time-consuming and the most comfortable way. But the situation is not so clear. For instance, research conducted by social media data has shown fatigue and anxiety of cross-border commuters (Chasset 2016). People stuck in traffic jam are frustrated, which cause decreased work productivity and less time for leisure. Even sleeping is of lower quality than required.

An interesting observation by a journalist Pete Hamill supports this development.

In fact, the automobile, which was hailed as a liberator of human beings early in this century, has become one of our jailers. The city air, harbour-cool and fresh at dawn, is a sewer by 10. The 40-hour week, for which so many good union people died, is now a joke; on an average day, a large number of people now spend three to four hours simply travelling to those eight-hour-a-day jobs, stalled on roads, idling at bridges or in tunnels. (Kane 1988)

Obviously, LSs cannot provide cleaner air. But, they may be applied in explaining behaviour of commuters and other journeys in cities. Mined knowledge from the data can be valuable input for motivation and optimization tasks.

Time is a valuable resource, but it is not effectively used. It is argued that self-driving cars will eliminate (or reduce) traffic accidents and improve efficiency. In addition, people will be able to focus on their work and spare time for leisure activities. But, it will not necessarily solve the problem of pollution in our car-oriented, unhealthy lifestyles (Moyser and Uffer 2016).

We should offer more healthy and efficient alternatives: public transport and cycling. The significant problem of the public transport lines is their organizing into the star structure, i.e. majority of lines have terminal stop in the city centre. Study in Aalborg has shown that people also commute between two more or less close points on periphery (Reinau 2016). Presumably, the same holds in other cities. This flow can be explained and verbalized by fuzzy cognitive maps. An example is verbalization of causalities in Fuzzy Cognitive Map (FCM) build on the taxi passengers flows by temporal protoforms of LSs is examined in Wehrle et al. (2016). In this direction, privacy issues should be seriously considered when personalized smart cards are used in public transport. The other option is anonymized smart cards provided by the local government for commuters to use them in ticket machines on entrance and exit.

Another alternative, a more healthy, is cycling. Nowadays, many bicyclists have in their possession smart phones. Study by Huang et al. (2016) has shown a possibility for motivation by sending information to bicyclists how much emissions they saved. Apparently, one bicyclist cannot save planet, but effect of millions cyclists is visible. Local authorities might provide tax benefits for cyclists, because cycling reduces needs for roads maintenance and improve health conditions. However, the approach of Huang et al. (2016) is not ideal for cyclists which concern their privacy. Malicious software might trace behaviour of cyclists, which is a valuable information

Table 6 Data collected from cyclists

Cyclist ID	Age	Length (km)	Day	Reason
c 1	25	7.37	1	Work
c 1	25	4.13	1	Shopping
c 1	25	11.02	2	Shopping
...				
c 2	21	8.32	1	Work
c 2	21	8.58	2	Work
c 2	21	2.44	2	Shopping
...				
c p	35	38.72	6	Recreation
c p	35	42.13	7	Recreation
c p	55	26.54	13	Recreation

for burglars, for example. The benefit is that local authority will be able to provide them reward for protecting environment and for saving local budget. Anyway, this is on cyclists to weight pros and cons. The option for non-registered cyclists is to send their daily routes by email. The website surveys could be also solution. Definitely, local authorities would have better overview of cycling in the city.

Having such data, we can built data warehouse similarly as was examined in Sect. 4.3.1 or FCMs as in Wehrle et al. (2016). Such data can be used for comparing public transport travellers, taxi passengers and cyclists. In this way routes or areas of significantly lower density of cyclists, but higher density of cars can be easily recognized and interpreted as, e.g. *Passengers rarely travel by bicycles from offices in area A to the railway station; Passengers often travel by taxis or cars from offices in area A to the railway station*. Further activities regarding new lanes for cyclists should be focused on such areas.

Let us have collected data in the structure shown in Table 6. From such table we can summarize variety of information for local authorities and planners as well as for citizens-cyclists.

Motivation in the sense of Huang et al. (2016) should be focused on working days. Peoples who are heavy users of cars during working days and heavy users of bicycles on weekends are not target group when we want to motivate commuting without cars.

Example 6 The task is creating suitable way for informing cyclists about their riding behaviour in comparison to average values. For this purpose illustrative data from Table 6 are aggregated and shown in Table 7. Let the average for month M is 11.23 km per day. Summary of structure *your length of ride is around average* is not informative enough. This summary is the same for person who use bicycle every day and for someone who rarely cycle on working days, but is heavy cycler during weekends. The better option are quantified linguistic summaries. For this purpose we have constructed five linguistic labels in a similar way as in Fig. 8, but with different number of labels and their parameters. These labels are: highly under average, slightly under average, around average, slightly above average and highly above average. All the membership functions are shown below:

$$\mu_{\text{highly under av}}(x) = \begin{cases} 1 & x \le 6 \\ \frac{8.5-x}{2.5}, & 6 < x < 8.5 \\ 0, & x \ge 8.5 \end{cases}$$

$$\mu_{\text{slightly under av}}(x) = \begin{cases} 0 & x \le 6 \vee \ge 11.23 \\ \frac{x-6}{2.5}, & 6 < x < 8.5 \\ 1, & 8.5 \ge x \le 10 \\ \frac{11.23-x}{1.23}, & 10 < x < 11.23 \end{cases}$$

$$\mu_{\text{around average}}(x) = \begin{cases} 0 & x \le 10 \vee \ge 12.5 \\ \frac{x-10}{1.23}, & 10 < x < 11.23 \\ 1, & x = 11.23 \\ \frac{12.5-x}{1.27}, & 11.23 < x < 12.5 \end{cases}$$

$$\mu_{\text{slightly above av}}(x) = \begin{cases} 0 & x \le 11.23 \vee \ge 15.5 \\ \frac{x-11.23}{1.27}, & 11.23 < x < 12.5 \\ 1, & 12.5 \ge x \le 14 \\ \frac{15.5-x}{1.27}, & 14 < x < 15.5 \end{cases}$$

$$\mu_{\text{highly above av}}(x) = \begin{cases} 0 & x \le 14 \\ \frac{15.5-x}{1.5}, & 14 < x < 15.5 \\ 1, & x \ge 15.5 \end{cases}$$

These membership functions are linear for the sake of simplicity and to ensure limited support for the inner terms (around average, slightly above average and slightly under average). The label *around average* can be expressed by function $\mu_{\text{around average}}(x) = (1 + (x - 11.23)^2)^{-1}$, but its support cover the whole domain, that is, matching degrees are greater than zero on the whole domain, although approaching zero for values far from 11.23.

Applying (1), (10) and relative fuzzy quantifiers from Fig. 1 with parameters $p_m = 0.6$ and $q_m = 0.85$ for quantifier *most of*, $p_h = 0.35$, $q_h = 0.45$, $u_h = 0.55$ and $z_h = 0.65$ for quantifier *about half* and parameters $u_f = 0.2$, $z_f = 0.4$ for quantifier *few*, we inform cyclist $c\ 1$ in the following way:

- few days your length of rides is slightly under average with validity 1
- about half of days your length of rides is about average with validity 0.971
- about half of days your length of rides is slightly above average with validity 0.222

We are now on the half way of explaining behaviour by an easily understandable way. Validities are not the best way for emphasizing sentences' relevance. We should add adjectives or highlight sentences by formatting. □

A suitable way for interpreting mined knowledge from the data is on dashboards, which provide collections of multiple visual components on a single window, so

Table 7 Aggregated data for cyclist *c 1*

Day	Length	Day	Length	Day	Length
1	12.507	11	10.793	21	12.312
2	11.023	12	12.982	22	12.776
3	11.479	13	11.803	23	11.936
4	11.293	14	11.282	24	9.212
5	12.120	15	11.159	25	11.416
6	12.403	16	12.039	26	11.327
7	10.656	17	11.652	27	12.003
8	11.234	18	11.008	28	12.646
9	11.320	19	11.224	29	12.554
10	11.265	20	11.215	30	11.854

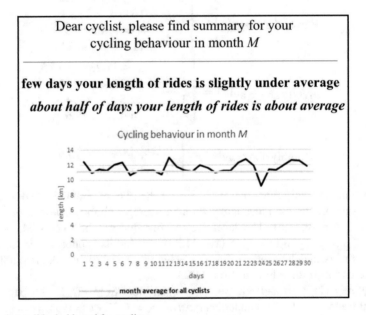

Fig. 9 A possible dashboard for cyclists

information is monitored at a glance (Few 2006). In Example 6 the structure of mined knowledge is realized by LSs. These summaries have different validities. Thus, we can adapt the following way: (i) sentences of validity equal to 1 are emphasized as bold, (ii) sentences bearing significant validity, e.g. $0.75 \leq v \leq 1$ are emphasized as, let say, bold-italic, (iii) sentences having medium validity are emphasized as italic. Other views on data can be graphically interpreted. A possible dashboard explaining data in Table 7 is plotted in Fig. 9, where LSs and graph explain time series for the selected month based on daily data.

We can create more sophisticated graphs and sentences. It is worth noting that we should not neglect the following facts: (i) detailed dashboard is less legible on smart phones displays, (ii) we want to provide relevant information to motivate cyclists, not to overwhelm them with data and information.

Aforementioned research on cross-border commuters by cars (Chasset 2016) has shown complaints of decreased quality of live. The same might appear for commuters inside large cities or in commuting from beyond city borders to the centre. These observations raise needs for measuring stress and other parameters (length of sleep, blood pleasure.....) on volunteers agreed to wear sensors. In this way, we could compare quality of live among cyclists, public transport users and car users. In addition, LSs can be converted into IF-THEN rules considering quality measures (Hudec 2017; Wu et al. 2010). For instance, if summary of structure *most of cyclists have high quality sleep* is recognized with validity 0.89 and for commuters by public transport with validity 0.78, then we can create the following weighted rules:

if commute by cycling, then quality of sleep is high [0.89]
if commute by public transport, then quality of sleep is high [0.78] following the structure in Wu et al. (2010) or

if commute by cycling, then quality of sleep is high with cf = 0.89
if commute by public transport, then quality of sleep is high with cf = 0.78
following the structure in Branco et al. (2005).

Such rules might be used as a further motivation for commuters to consider cycling and public transport as better alternatives and for further research focused on e.g. health conditions.

Commuting by cars might be still the best option in parts of cities, which are inefficiently covered by public transport, or where cycling lanes are missing. FCM of cars, public transport, taxi passengers and cyclists might express these behaviours (how they differ and overlap). Smart phones with GPS and smart public transport cards are the easiest options for data collection, but traditional surveys should be also applicable in collecting data.

4.4 Tourism Behaviour in Summaries

Nowadays, almost all travellers are owners of mobile phones. Open borders in Schengen Area and alternative accommodations cause significant difference between official statistics (data from registered accommodation units - number of guests and length of stays by countries) and number of foreign visitors gathered from the anonymized mobile positioning data (Ahas et al. 2011). On the city level we might expect the same and presumably the bigger and more attractive city, the higher difference.

Mobile positioning data have shown their potential in supporting official statistics, mainly in the tourism area (Demunter and Reis 2015). In this field, variety of data analyses, visualization techniques, and protecting sensitive data have been conducted (Ahas et al. 2011), (Ahas et al. 2010; Tiru 2014).

These studies were mainly focused on the countries levels, but cities might also benefit from such studies. Local authorities (e.g. department for tourism), small businesses in tourism, taxi companies, marketers could prosper analysing such data. These data belong to the category so-called big-data due to the large appearance of CDR (Call Data Recording) and GPS data. Mobile operators' data are of a high quality, because billing is based on them. The inspiration is cooperation in Estonia between national statistical office, research institutions and mobile operators who provide anonymized data.

The following question is also relevant: How to interpret knowledge in these data for different categories of users? Variety of data analyses and visualization techniques have been developed in this field. However, the efficient way is not always realizable through tables, graphs and maps, especially when we cope with semantic uncertainties such as *frequent, occasionally, remote country, short stay, most of* and different levels of statistical literacy among stakeholders. An alternative is interpretation by short quantified sentences of natural language. Stakeholders in a city may be interested to see whether a particular sentence such as *most of visits from middle distanced countries entered on airports have short stay, few long stays are from low income countries* and *most of visits use near by airport as an entrance hub to the whole country* holds.

These data can be migrated into the data warehouses for further analyses. Dimensions are time, visited location and country of foreign visitors' origin. In this way, we can focus our interest on locations inside city (slice, drill down and dice multidimensional data structures). Moreover, we can include attributes such as distance, GDP and other relevant ones into the dimensional table related to visitors' country of origin.

Example 7 A city tourism agency wishes to know usual length of stay for visitors from the EU countries. Hence, the term set for summarizer is $\overline{S}$ = {short, medium, long}. Similarly, the term set for relative quantifier is $\overline{Q}$ = {few, about half, most of}. According to (1) and (10) we should evaluate the following nine sentences:

- most of visits from EU countries are of short stay
- most of visits from EU countries are of medium stay
- most of visits from EU countries are of long stay
- about half visits from EU countries are of short stay
- about half visits from EU countries are of medium stay
- about half visits from EU countries are of long stay
- few visits from EU countries are of short stay
- few visits from EU countries are of medium stay
- few visits from EU countries are of long stay

Belonging to the EU is a sharp restriction. Thus, we use the basic structure of LS (1). The solution can be expressed in a similar way as in Table 1.

In addition, we can drill down to the visitors countries level. These summaries can be expressed on thematic maps by hues of selected colour following the validities for particular countries. For these summaries quality measure is required. When very small number of visits from a particular country is recorded, then we cannot create reliable behaviour patterns for visitors from this country. □

Another possibility is examining development over time. Let us have localities inside city like castles and museums, where higher interest of tourists is assumed. If visitors are asked to declare country of origin while buying tickets, then we know the distribution of countries. But, what we do not know is the ratio of sold tickets to the total number of foreigners wandering around these locations. For example, the low ratio might indicate problems such as difficulties to find entrance, high price and the lack of information in foreign languages. If ratio of visitors from several countries significantly vary, then analysts in tourism agency should find reasons for such behaviour and suggest improvements. CDR data cover foreign phone numbers, which were active around localities of interest. Usually people inform their friends and relatives by calls or messages about situation around the locality or sent post on social networks. This task can be solved in the similar way as in Example 5 by adjusting membership functions of linguistic terms shown in Fig. 8.

Accommodation units have duty to provide data regarding number of guests and length of stay by countries to the national statistical offices. Let us imagine that the number of accommodated foreigners by small areas is available in real time. If dispatcher of a taxi company know that majority of visitors in area A are from countries speaking language L, then he might sent driver who has language skills of L. Taxi driver and passenger could have better feeling during the journey.

Although, in this section the focus is on the mobile positioning data, privacy of visitors should not be violated. Mobile operators collect CDR data, otherwise they will not be able to provide roaming services and generate bills. The only way to avoid this is to travel without mobile phone or do not use it actively. Operators, if agree to cooperate, should provide anonymized data in the same way as accommodation units are doing for a long period: number of visitors and length of stay (possibly also whether visit is repeated).

In addition, on entrance and exit hubs in cities (railway station, airport, bus terminal) we can often find panels with several buttons indicating level of satisfaction (very low, low, medium, high, very high). Visitors should simply press one of them. It gives space for applying protoforms explained in Sect. 3. Summaries such as *most of unsatisfied visitors are in early hours*, *regularly on Sundays evening satisfaction is low*, *at least about half visitors is satisfied in rush hours* and the like explain behaviour in a human friendly way. In these summaries we recognize vague concepts: regularly, early hours, rush hours. Although these concepts do not have sharp boundaries, their meaning is clear.

4.5   Estimation of Missing Values

This problem appears when values are not known for some observations. Missing values appear due to several reasons (e.g. technical problems with sensors, non-availability of instruments to measure phenomena in all locations and reluctance of citizens to cooperate in surveys). This problem influences computing aggregations and mining information, among others. This is another topic where acquired knowledge in official statistics might be beneficial, because these institutions cope with data imputation regularly.

One of used methods is the Hot Deck imputation due to its simplicity and good results, even though its theory is not as developed as for the other methods (Andridge and Little 2010). Soft computing approaches: neural networks (Juriová 2011) and genetic programming (Kľučik 2012) are also valuable contribution in this field, though somewhat complex. Their advantage is in estimating values of datasets containing attributes which are complexly influenced and might significantly vary due to internal and external influences. On the other hand, LSs are not so complex. They are able to process non-linear and flexible, but relatively stable relations. For example, if a fill-level sensor on bin does not operate, we can estimate its value from sensors covering similar locations for similar households about the same time.

When validity is significant but not sufficiently high, or quality is insufficient, presumably in a more restricted part of a database validity is stronger. This can be realized by adding semantically reasonable attribute(s) into the restriction part of LS as is illustrated in Fig. 10.

For instance, when summary *most of small households have small amount of waste* has insufficient validity or data coverage, then sentence *most of small households and in small income area have small amount of waste* might have higher validity (2) and quality (3). If this summary is valid, then we can roughly estimate missing values for households belonging to this category.

4.6   Data Exchange, Disclosure, Computational Effort and Data Standards

Data privacy is a sensitive topic in any field dealing with the personal data and in avoiding data disclosure when sharing and publishing aggregated data.

The success of the cognitive city rests on the principle, known since first civilizations: data and information sharing among active citizens. Nowadays, we use powerful tools for sharing, keeping (literally forever) and processing data. It raises the issue of data privacy and security to the new level. On the other hand, when only 1–2% of the urban population is willing to play an active role in this direction, the implications on improvements would be dramatic (Mostashari et al. 2011).

This chapter has emphasized that the contribution to cognitive cities can be realised without keeping track about citizens everyday activities by means of smart phones

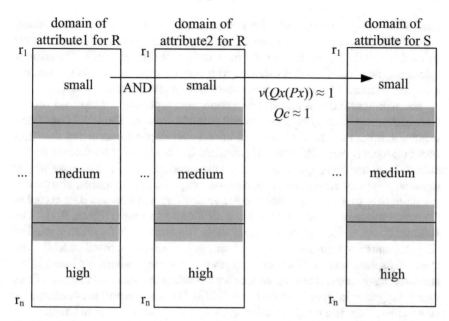

Fig. 10 LS with restriction part consisted of two attributes merged with the *and* connective, adjusted form Hudec (2014)

with GPS, sensors and the like. If low percentage of citizens voluntary agree to play an active role and other are able to participate anonymously by surveys, we can reach new quality and satisfaction on both sides: citizens and local government. In this direction, experiences from national statistical institutes regarding surveys, data imputation, data disclosure problems, data exchange standards and motivation are valuable. In addition, data collected within cities might be valuable data source for national statistical institutes. Apparently, in modern world overwhelmed with data, cooperation between statistical offices and cities could be beneficial for both parties.

Data from sensors measuring pollution or traffic density should be always at disposal. The same holds for aggregated data, when the data disclosure is not the problem. LSs provide summarized or aggregated explanation. Thus, risk of data disclosure is minimized, especially when summarizing from the large data sets. The promising way is comparing human opinions with sensor data to see how correlate these opinions with the objectively measured data.

Computational effort is an significant issue when we would like to efficiently and quickly reply to citizens, researchers, governance and journalists questions. When we want to calculate summaries among attributes, it is necessary to compute high number of membership degrees (Niewiadomski et al. 2006). Hence, the computation might take much time and might be costly. Practice and experiments have shown that only a subset of tuples partially or fully satisfies summarizer and restriction and usually several summaries in (10) convey high validity. By the SQL-like queries we compute proportions (1), (2) only for tuples, which at least partially belong to summarizer and restriction. Computation of validities for summaries with the expected low validity is pointless. Hence, proportions and validities of already processed summaries can be used for excluding irrelevant ones from the queue (Hudec 2016). The effort in calculation and optimization is balanced with answers expressed by short quantified sentences of natural language understandable for all stakeholders.

The significant issue is efficient data collection, especially for repeated surveys with slight modifications in questions and possible answers. The DDI (Data Documentation Initiative) standard has been initially developed for archiving data from social sciences surveys. Recently, this standard is extended to cover fields from data collection to archiving (Thomas 2015). This standard contains hierarchy of tags explaining type of survey, methodology, version, who is responsible, structure of questionnaire, particular questions, possible answers and how are data recorded. The same structure can be used for data collection form sensors (Block 2011). This standard however does not handle fuzzy data. The same holds for another standard SDMX (Statistical Data and Metadata eXchange), which is also based on XML. This standard facilitates the exchange of aggregated data among national and international statistical institutions. Recently, an idea for extending this standard to handle fuzzy data is suggested in Hudec and Praženka (2016). If such standards are beneficial for smart cities, there is a perspective for applying them and adjusting to handle fuzzy data.

Usually smartness of city is considered inside a single city. Focus is on approaches which improve living conditions. But, what could be valuable is comparison among cities. Not for pure competition but to exchange experiences and achievements. Summaries are aggregations and therefore sensitive micro data are not exchanged.

5 Conclusion

The shift towards smart and further to smarter or cognitive cities has already started. Technology companies have created a variety of technical solutions specially designed for smart cities, or solutions which can be applied for increasing smartness such as selfdriving cars, navigation, tools for optimizing leisure time, and body sensors. We should be careful because these solutions will not necessarily solve our unhealthy lifestyles and also might cause that citizens in cognitive cities become less cognitive. Too much relying on autonomous car parking systems, navigation to the destination point and similar gadgets might cause people to loss their natural skills like orientation is space.

The second issue is a demanding task regarding keeping sensitive personal data confidential and undisclosed. On the other hand, cognitivity in cities can be improved by collecting and analysing non-confidential data and working with aggregated or summarized data. Moreover, such data can be exchanged among institutions inside a city or even among cities to compare developments, for example.

In all these tasks, we should not neglect the semantic uncertainty in the data and fuzziness in the concepts. This chapter has discussed perspectives in summarizing from the data by short sentences of natural language. This approach is able to summarize from a variety of data types in a comprehensible way for all stakeholders, including those with lower level of statistical literacy and skills with information technologies. We should not forget visually impaired and colour-blind citizens, which might benefit from less complex data interpretation, whereas for some of them textual interpretation (e.g. linguistic summaries) is the most suitable, because it can be interpreted by a text-to-speech synthesis system. In addition, we should not forget that cognitive approaches in cities need to be beneficial for all citizens not only for technology enthusiasts.

This approach can fulfil citizens' needs to be efficiently informed. By linguistic summaries, citizens can examine achievements more easily, recognize reasons for new regulations and are better prepared for voting. Citizens' opinions is a relevant source for local administration, but these opinions cannot always be expressed without vagueness. Our approach provides a framework for managing this uncertainty in data collection and summarization. Local authorities, planners, marketers, public relation units get more legible overview of citizens' opinions, concerns and needs. The same approach is able to summarize from sensor data and data regarding the behaviour of tourists. Converting collected data into the data warehouse schemes provides a framework for answering variety of questions asked by local authorities, dispatchers, citizens and journalists.

Examples are realized on the illustrative data in order to show the benefits and open further discussions related to possibilities, strengths and weaknesses, adopting to particular needs, and developing full-featured software tools for the benefit of all stakeholders.

References

Ahas R, Aasa A, Roose A, Mark U, Silm S (2010) Evaluating passive mobile positioning data for tourism surveys: an estonian case study. Tour Manag 29:469–486

Ahas R, Tiru M, Saluveer E, Demunter C (2011) Mobile telephones and mobile positioning data as source for statistics: Estonian experiences. In: New techniques and technologies for statistics (NTTS 2011). Brussels

Almeida R, Lesot M-J, Bouchon-Meunier B, Kaymak U, Moyse G (2013) Linguistic summaries of categorical time series patient data. In: 2013 IEEE international conference on fuzzy systems (FUZZ-IEEE 2013). Hyderabad, pp 1–8

Andridge R, Little R (2010) A review of hot deck imputation for survey non-response. Int Stat Rev 78:40–64

Arguelles L, Triviño G (2013) I-struve: automatic linguistic descriptions of visual double stars. Eng Appl Artif Intell 26:2083–2092

Beliakov G, Pradera A, Calvo T (2007) Aggregation functions: a guide for practitioners. Springer, Berlin

Block W (2011) Documenting a wider variety of data using the data documentation initiative 3.1. DDI Working Paper Series-Longitudinal Best Practice, No 1. DDI alliance

Bojadziev G, Bojadziev M (2007) Fuzzy logic for business, finance and management, 2nd edn. World Scientific Publishing Co, London

Boran F, Akay D, Yager R (2016) An overview of methods for linguistic summarization with fuzzy sets. Expert Syst Appl 61:356–377

Branco A, Evsukoff A, Ebecken N (2005) Generating fuzzy queries from weighted fuzzy classifier rules. In: ICDM workshop on computational intelligence in data mining. Houston, pp 21–28

Campbell M, Swinscow T (2009) Statistics at square one, 11th edn. Wiley, West Sussex

Chasset P (2016) Tracking cross–borders commuters fatigue through twitter: a new sampling strategy. In: Mobile Tartu 2016. Tartu

Date C (2006) Date on databases: writings 2000–2006. Apress, New York

Demunter C, Reis F (2015) Using mobile positioning data for official statistics: daydream nation or promised land? In: New techniques and technologies for statistics (NTTS 2015). Brussels

Doukas C, Maglogiannis I (2012) Bringing iot and cloud computing towards pervasive healthcare. In: Sixth international conference on innovative mobile and internet services in ubiquitous computing (IMIS 2012). Palermo, pp 922–926

Duraj A, Szczepaniak PS, Ochelska-Mierzejewska J (2016) Detection of outlier information using linguistic summarization. In: Andreasen T, Christiansen H, Kacprzyk J, Larsen H, Pasi G, Pivert O, De Tré, G, Vila MA, Yazici A, Zadrożny, S (eds) Flexible query answering systems 2015. Springer, International Publishing Switzerland, pp 101–113

Few S (2006) Information Dashboard Design. O'Reilly, North Sebastopol

Finger M, Portmann E (2016) What are cognitive cities? In: Finger M, Portmann E (eds) Towards cognitive cities-advances in cognitive computing and its application to the governance of large urban systems. Springer, International Publishing Switzerland, pp 1–11

Galindo J, Urrutia A, Piattini M (2006) Fuzzy databases-modeling, design and implementation. Idea Group Publishing, Hershey

Giesen DE (2011) Response burden in official business surveys: measurement and reduction practices of national statistical institute. Deliverable 2.2, Blue-Ets Project. European Commission, Brussels

Hirota K, Pedrycz W (1999) Fuzzy computing for data mining. Proc IEEE 87:1575–1600

Huang A, Fioreze T, Thomas T (2016) On personalised positive incentives to reduce car use: a micro–experiment to promote cycling on the campus of the University of twente. In: Mobile Tartu 2016. Tartu

Hudec M (2014) Linguistic summaries applied on statistics-case of municipal statistics. Austrian J Stat 43:63–75

Hudec M (2015) Storing and analysing fuzzy data from surveys by relational databases and fuzzy logic approaches. In: XXV IEEE international conference on information, communication and automation technologies (ICAT 2015). Sarajevo, pp 220–225

Hudec M (2016) Fuzziness in information systems –How to deal with crisp and fuzzy data in selection, classification and summarization. Springer, International Publishing Switzerland

Hudec M (2017) Merging validity and coverage for measuring quality of data summaries. In: Kulczycki P, Kóczy L, Mesiar R, Kacprzyk J (eds) Information technology and computational physics, advances in intelligent systems and computing. Springer, International Publishing Switzerland, pp 71–85

Hudec M, Praženka D (2016) Collecting, storing and managing fuzzy data in statistical relational databases. Stat J IAOS 32:245–255

Hudec M, Torres Van Grinsven V (2013) Business' participants motivation in official surveys by fuzzy logic. In: 1st Eurasian multidisciplinary forum (EMF 2013), vol 3. Tbilisi, pp 42–52

Hudec M, Vučetić M, Vujošević M (2014) Synergy of linguistic summaries and fuzzy functional dependencies for mining knowledge in the data. In: 18th IEEE International conference on system theory, control and computing (ICSTCC 2014). Sinaia, pp 335–340

Jensen C, Pedersen T, Thomsen C (2010) Multidimensional databases and data warehousing. Morgan & Claypool Publishers

Juriová J (2011) Use of neural networks for data mining in official statistics. In: New techniques and technologies for statistics (NTTS 2011). Brussels

Kacprzyk J, Wilbik A, Zadrożny S (2008) Linguistic summarization of time series using a fuzzy quantifier driven aggregation. Fuzzy Sets Syst 159:1485–1499

Kaltenrieder P, D'Onofrio S, Portmann E (2015) Enhancing multidirectional communication for cognitive cities. In: Second IEEE international conference eDemocracy & eGovernment (ICEDEG 2015). Quito, pp 38–43

Kane T (1988) The new Oxford guide to writing. Oxford University Press, Oxford

Keefe R (2000) Theories of vagueness. Cambridge University Press, Cambridge

Kimball R, Ross M (2013) The data warehouse toolkit, 3rd edn. Wiley, New York

Klučik M (2012) Estimates of foreign trade using genetic programming In: 46th Scientific meeting of the Italian statistical society (SIS 2012). Rome

Ladner A, Meier A (2014) Digitale politische partizipation-spannungsfeld zwischen mypolitics und ourpolitics. HMD Praxis der Wirtschaftsinformatik 51:867–882

Lesot M-J, Moyse G, Bouchon-Meunier B (2016) Interpretability of fuzzy linguistic summaries. Fuzzy Sets Syst 292:307–317

Li Q (2013) A novel likert scale based on fuzzy set theory. Expert Syst Appl 40:1609–1618

Likert R (1932) A technique for te measurement of attitudes. Arch. Psychol. 22(140):1–55

Liu B (2011) Uncertain logic for modeling human language. J Uncertain Syst 5:3–20

Miller G (1956) The magical number seven, plus or minus two. Some limits on our capacity for processing information. Psychol Rev 63:81–97

Mostashari A, Arnold F, Mansouri M, Finger M (2011) Cognitive cities and intelligent urban governance. Netw Ind Q 13:4–7

Moyse G, Lesot M-J, Bouchon-Meunier B (2013) Linguistic summaries for periodicity detection based on mathematical morphology. In: 2013 IEEE symposium on foundations of computational intelligence (FOCI). Singapore, pp 106–113

Moyser R, Uffer S (2016) From smart to cognitive: a roadmap for the adoption of technology in cities. In: Finger M, Portmann E (eds) Towards cognitive cities-advances in cognitive computing and its application to the governance of large urban systems. Springer, International Publishing Switzerland, pp 13–35

Niewiadomski A (2002) Appliance of fuzzy relations for text documents comparing. In 6th Conference on neural networks and soft computing (ICNNSC 2002). Zakopane, pp 347–352

Niewiadomski A, Ochelska J, Szczepaniak P (2006) Interval-valued linguistic summaries of databases. Control Cybern 35:415–443

OECD (2012) OECD environmental outlook to 2050. The consequences of inaction. OECD Publishing, Paris

Ovaska S (2005) Introduction to fusion of soft computing and hard computing. In Ovaska S. (ed) Computationally intelligent hybrid systems: the fusion of soft computing and hard computing. Wiley, Hoboken, pp 5–30

Perticone V, Tabacchi M (2016) Towards the improvement of citizen communication through computational intelligence. In: Finger M, Portmann E (eds) Towards cognitive cities-advances in cognitive computing and its application to the governance of large urban systems. Springer, International Publishing Switzerland, pp 83–100

Rasmussen D, Yager R (1997) Summary sql-a fuzzy tool for data mining. Intell Data Anal 1:49–58

Reinau K (2016) Smart card data in a peripheral region: modularity analysis and the geography of passenger flows. In: Mobile Tartu 2016. Tartu

Ruspini E (1969) A new approach to clustering. Inf Contr 15:22–32

Smits G, Pivert O, Girault T (2013) Reqflex: fuzzy queries for everyone. In: 39th international conference on very large data bases. Trento, pp 1206–1209

Terán L, Kaskina A, Meier A (2016) Maturity model for cognitive cities. In: Finger M, Portmann E (eds) Towards cognitive cities-advances in cognitive computing and its application to the governance of large urban systems. Springer, International Publishing Switzerland, pp 37–59

Thomas W (2015) Data capture: tracking data from source to results. In: North American DDI user conference (NADDI). Wisconsin

Tiru M (2014) Overview of the sources and challenges of mobile positioning data for statistics. In: International conference on Big data for official statistics, Beijing

Torres van Grinsven V (2015) Motivation in business survey response behavior. PhD thesis, University of Utrecht, Utrecht

Torres van Grinsven V, Snijkers G (2015) Sentiments and perceptions of business respondents on social media: an exploratory analysis. J Official Stat 31:283–304

Tudorie C (2008) Qualifying objects in classical relational database querying. In: Galindo J (ed) Handbook of research on fuzzy information processing in databases. Information Science Reference, Hershey, pp 218–245

van der Heide A, Trivino G (2009) Automatically generated linguistic summaries of energy consumption data. In: Intelligent system design and applications (ISDA 2009). Pisa, pp 553–559

Viertl R (2011) Fuzzy data and information systems. In: 15th WSEAS international conference on systems. Corfu, pp 83–85

Škrbić S, Racković M, Takači A (2011) Towards the methodology for development of fuzzy relational database applications. Comput Sci Inf Syst 8:27–40

Wehrle M, Osswald M, Portmann E (2016) Verbalization of dependencies between concepts built through fuzzy cognitive maps. In: Finger M, Portmann E (eds) Towards cognitive cities-advances in cognitive computing and its application to the governance of large urban systems. Springer, International Publishing Switzerland, pp 123–144

Wilbik A, Keller J (2013) A fuzzy measure similarity between sets of linguistic summaries. IEEE Trans Fuzzy Syst 21:183–189

Wilbik A, Keller J, Alexander G (2011) Linguistic summarization of sensor data for eldercare. In: 2011 IEEE international conference on systems, man, and cybernetics. Anchorage, pp 2595–2599

Wu D, Mendel J, Joo J (2010) Linguistic summarization using if-then rules. In: 2010 IEEE international conference on fuzzy systems. Barcelona, pp 1–8

Yager R (1982) A new approach to the summarization of data. Inf Sci 28:69–86

Yager R, Ford M, Canas A (1990) An approach to the linguistic summarization of data. In: 3rd International conference of information processing and management of uncertainty in knowledge-based systems (IPMU 1990). Paris, pp 456–468

Zadeh L (1996) Fuzzy logic = computing with words. IEEE Trans Fuzzy Syst 4:103–111

Zadeh L (2012) Computing with words-principal concepts and ideas. Springer, Berlin Heidelberg

Zimmermann H (2001) Fuzzy set theory-and its applications. Kluwer Academic Publishers, Dordrecht

Miroslav Hudec is currently Associate Professor at the University of Belgrade, Faculty of Organizational Sciences, Serbia. He received a Master's and PhD degree from the University of Belgrade on Operations Research. His work mainly focuses on fuzzy logic, knowledge discovery, and information systems. Miroslav Hudec is author or co-author of approximately 45 scientific articles and three books. He is a member of program committees of several international conferences, serves as an editorial board member of the Journal Applied Soft Computing and gave four invited talks. He was the representative of Slovakia on UNECE/Eurostat/OECD Meeting on the Management of Statistical Information Systems (2005–2013). He also participated in several international and national projects including FP7 Blue-ETS projects, where he also managed research activities.

An Exploration of Creative Reasoning

Enric Trillas, Sara D'Onofrio and Edy Portmann

Abstract In pursuit of advancing cognitive cities, this article introduces a first model of creative reasoning in a naïve way. Based on ordinary reasoning, by presenting formal deduction, a mathematical structure without too many formal constraints to allow 'creative jumps' in reasoning is elaborated. Facets of natural language as well as of human thinking are mentioned to emphasize the importance of creativity in human life as well as in cities, considering the existing imprecision and uncertainty in natural language. Therefore, this article gives the first hints of a possible mathematical model to enable creative reasoning (e.g., in cognitive cities).

Keywords Cognitive city · Conjecture · Connectivism · Fuzzy logic
Ordinary reasoning · Smart city

1 Introduction

> What requires creativity is not to transform the sun into a yellow spot, but a yellow spot into the sun.
>
> Pablo Picasso

1.1 Creativity is not seen as a miracle, but as a natural phenomenon. It can be only comprehended through the human brain functioning; namely through thinking. Creativity is broader than reasoning. Since finding something new is attributed to reasoning processes, it seems reasonable to search for the kind of reasoning being

E. Trillas
University of Oviedo, Oviedo, Spain

S. D'Onofrio (✉) · E. Portmann
Human-IST Institute, University of Fribourg, Boulevard de Pérolles 90, 1700 Fribourg, Switzerland
e-mail: sara.donofrio@unifr.ch

E. Portmann
e-mail: edy.portmann@unifr.ch; edy.portmann@gmail.com

© Springer Nature Switzerland AG 2019
E. Portmann et al. (eds.), *Designing Cognitive Cities*, Studies in Systems, Decision and Control 176, https://doi.org/10.1007/978-3-030-00317-3_4

able to allow creative results. Creativity is, and always has been, considered at the top of human intellectual activity and related to what in ordinary human reasoning can be called 'creative jumps' and 'dangerous shortcuts'. In some sense, creativity is reached through a type of reasoning opposed to formal deduction, which is supposed to be performed with no jumps and no shortcuts. Deductive reasoning consists of chains of (partial) reasoning whose steps are perfectly linked (deductively), each one with the former, without any jump in between. Creativity proceeds through 'creative reasoning' that allows innovation and can be seen as the greatest expression of human rationality. According to David Bohm's ideas (Bohm 2004), creativity can only truly flourish upon ceasing to live and reason mechanically. More often than not, creativity means a 'cut' in the corresponding field from which things become radically different.

Formal deduction is essentially a special type of ordinary deductive reasoning that, done in a formalized setting, starts from precisely established premises and ends with conclusions (called consequences) but is, notwithstanding, unable to reach anything not implicit or hidden in the premises. Formal deduction is performed with the 'calculus' in the framework in which the (external) reasoning is mimicked. In addition, in formal deduction, there are never contradictions among the consequences, something not guaranteed in ordinary reasoning. In the ordinary forms of deducing in plain language, it is not assured that all the consequences can satisfy Tarski's consequence operator (cf. Tarski 1956), as in formal deduction, whereas this operator is considered a good model of formal deduction. Hence, ordinary deduction should be distinguished from formal deduction; the second can be viewed as a 'mechanical' reasoning, something not to be confused with an 'easy' reasoning, as shown by the complex proofs of some mathematical theorems such as, for instance, Andrew Wiles's proof of Fermat's last theorem (cf. Wiles 1995).

By formal deduction, it is impossible to reach conclusions that are not implicit in the premises; it only facilitates what follows from these premises. It can be said that formal deduction is the deduction performed with 'paper and pencil' that can be reviewed step by step by another person. Creative reasoning, however, makes it possible to reach conclusions, regardless of the contents of the premises, by performing inference in a jump-like way without considering the formal order of deduction. Therefore, formal deduction and creative reasoning can be seen as the pearls of reasoning that nevertheless are not limited to these types and are interwoven with plain language.

To examine all types of reasoning, it seems strictly necessary to escape from excessively structured formal frameworks and the known logic methodologies to consider the mathematically unstructured setting of ordinary, everyday, or common-sense reasoning, in which the laws supposed in formal frames are not always valid (Trillas 2017). For instance, the linguistic particle 'and' does not always follow the commutative law; since time often intervenes in a telling, for instance, the meaning of '*She enters the room, and cries*' cannot be confused with the meaning of '*She cries, and enters the room*', which describes a different situation. Analogously, the associative law cannot be always presumed because it requires ignoring commas and can lead to misunderstanding linguistic statements.

In modeling the creative forms of reasoning, the problem can be described as extending the informational contents of the premises, the starting information statements, without too many formal constraints; trying to model 'something new' within a softer structure than is possible in mathematics. The underlying idea is to be as free as possible from formal constraints since, in principle, plain language appears unstructured. Of course, to perform such reasoning in a symbolic form, some weak mathematical framework is necessary; in particular, for analyzing ordinary reasoning through the Karl Menger's 'Exact Thinking' (cf. Menger 1974).

1.2 In the words of Sternberg (2003, p 89):

> Creativity is the ability to produce work that is novel (that is, original, unexpected), high in quality, and appropriate (that is, useful, meets task constraints). Creativity is a topic of wide scope, important at both the individual and the societal levels for a wide range of tasks domains,

and also,

> The economic importance of creativity is clear because new products or services create jobs. Furthermore, individual organizations, and societies must adapt existing resources to changing tasks demands to remain competitive.

Nothing else is necessary to capture the relevance of creativity for the concerns of complex intelligent systems, in particular, the ones currently known as smart or cognitive cities. Cities that are using information and communication technologies to access, process and use urban data to develop the city in a more social, ecological and efficient way are called smart cities (Portmann and Finger 2015). By complementing the smart technologies with soft computing methods (cf. D'Onofrio and Portmann 2015) and by enriching them with learning algorithms, a common urban knowledge base can be achieved (D'Onofrio and Portmann 2017). Thus, a city should count on highly creative people, especially on research centers with a true creative reputation; centers devoted to science, technology, arts, and so on, to advance the city's processes creatively. Today, not only artistic creativity is needed but also creativity in business, craftsmanship, literary, scientific, philosophical, medical, and entrepreneurial activities, etc. By letting go the formal constraints of reasoning, the city with its citizens can creatively build an urban intelligence (i.e., a collective intelligence in a city) and evolve into a cognitive city (D'Onofrio and Portmann 2017; Moyser and Uffer 2016). It is essential that the reasoning process a city is using and will use in the future is like the way its citizens think. Therefore, creativity is a crucial feature in a city that cannot be ignored: no true cognitive city, nor even a smart one, can exist without large doses of relevant creativity, and better if it is driven by a 'creative educational system' like the one advocated by Robinson and Aronica (2016). The development of cognitive cities is a great challenge for the future of mankind, and it is essential to cultivate in them an environment full of creativity.

1.3 Some steps towards a study of such creative reasoning without excessively formal constraints are currently possible, and this article tries to present a naïve reflection on them from some previously introduced patterns on ordinary reasoning (cf. Trillas 2015; Trillas and Eciolaza 2015; Trillas 2017). It tries to offer an analysis through some clues towards the challenge of its possible formalization in mathematical terms with a general enough character to allow its application to particular types of reasoning. The study of creativity, begun in psychology and philosophy and now continued in the field of neuroscience, does not yet include a mathematical model able to show how creative reasoning can be distinguished within reasoning; this paper tries to offer some hints toward such a possible model.

From a purely biological perspective, reasoning and creativity consist of neurological processes in the brain; hence, it will not be scientifically well understood until the neurosciences describe how the brain actually functions. Nevertheless, meanwhile, to capture a possible mechanism for intellectual life, even in a provisional form, this model at least offers the suggestive beauty of its theoretical simplicity for representing the whole phenomenon of reasoning and, within it, distinguishing creative reasoning.

1.4 This article is structured as follows: Sect. 2 explains ordinary reasoning, while Sect. 3 introduces a first naïve way of creative reasoning with the aim of giving a basic understanding of both approaches to reasoning. Section 4 gives further notes about reasoning in general, and Sect. 5 concludes this article with insight into further research. The exploration of creative reasoning seeks to foster cognitive cities.

2   Ordinary Reasoning

2.1 The patterns cited in 1.3 are based on the linguistic relationship

'q does not follow from p',

symbolically denoted by $p \nleq q$, for meaningful statements p and q. This symbolically represented relationship is not formally defined but presumed to be primitive, recognizable thanks to the human capability of understanding linguistically articulated reasoning as a manifestation of thinking; it is just taken as the primary spring of reasoning, involving both the terms 'follow' and 'not', supposedly to be acquired during the process through which people simultaneously learn both language and reasoning. A simple example involves the statements

p = 'The weather today is very hot', and q = 'I will wear an overcoat',

since everybody will accept $p \nleq q$, that if p then it cannot be necessarily presumed that q. Denoting the linguistic negation 'not p' by p', then $p \nleq q$', symbolizing

'from p, it does not follow not q',

opens the window to consider a linguistic conjecture, or guessing, of q, from what p expresses (i.e., known information on something). For instance,

q = 'It will rain',

is a guess based on

$$p = \text{'There are very dense clouds'},$$

since everybody will accept $p \nless q'$:

$$q' = \text{'It will not rain'},$$

that not-q does not follow from knowing p.

Finally, and provided that if $p < q$, q 'follows' from p, q is an informal, linguistic, or ordinary, consequence of p, and if $p < q'$ then q refutes p, or that q is a refutation of p. A reason for beginning with $\nless$ instead of with $<$, is that, often, it is easier to recognize when q does not follow from p than when q does follow from p.

It can be questioned if and when an ordinary consequence of p is also a conjecture of p; when this 'consequencing' or ordinary deduction is a particular case of guessing. An answer to this question can be obtained under the following three hypotheses:

(1) The negation (') reverses the relation $<$, that is, provided $p < q$, then $q' < p'$;
(2) The premise p is not self-contradictory, that is, it cannot be true that $p < p'$;
(3) p cannot refute itself. Self-contradiction can be viewed as the worst sin of reasoning, since everybody will refuse a statement refuting itself;
(4) The triplet (p, q', p') is $<$-transitive, that is, provided $p < q'$, and $q' < p'$, then $p < p'$.

Indeed, if $p < q$, and if $p < q'$, then from $q' < p'$ will follow the absurd consequence $p < p'$; hence, it should be true that $p \nless q'$; consequence q is a conjecture of p.

In addition, under the same former hypotheses, there cannot exist two contradictory consequences of p; indeed, if $p < q$, $p < r$, and $q < r'$, it follows that $p < r'$ and $r' < p'$; thus, $p < p'$. Based on these presumptions, no consequence can refute other consequences.

Conditions (1) and (2) seem reasonably weak and not too debatable, but hypothesis (3) on 'transitivity' is actually a strong one, since there are many situations in which $<$-transitivity is not clearly presumable. For instance, in winter and with

$$p = \text{'There are very dense clouds, it is not windy, and the temperature is very low'},$$

it can be accepted

$$q = \text{'It will snow'} \ (p < q);$$

and from q it can be accepted that

$$r = \text{'The streets will become slippery'} \ (q < r);$$

but it is not immediately supposable that $p < r$. The triplet (p, q, r) is not transitive, and what everybody can 'just guess' is $p \nless r'$, that streets will become slippery; that is, conjecturing such a possibility. Notice that, instead, it can be accepted that

$$p \cdot q \, < \, r \, (r \, \text{follows from 'p and q'})$$

The triplet's transitivity seems to be linked solely to what concerns consequencing, and its lack could limit people to conjecturing; it seems that this lack produces the holes that can only be escaped by jumping, by making a non-deductive guess.

Like Henry Poincaré, who qualified transitivity as a hypothesis added to physics by its mathematical models and imagined a physical continuum without it but adapted to common experience (Poincaré 2011), a linguistic type of 'reasoning continuum' can be presumed in which transitivity is a 'local property', holding only for some particular triplets of statements, allowing ordinary deduction.

Another example of such linguistic continua is given by chains of synonyms; each term is a synonym of the former, but the final term is not a synonym of the first term; synonymy is not transmitted along the chain. Such continua, obviously, have holes that, when essential, must be passed using 'creativity'; something people do that is very difficult for computers, whose functioning is defined by algorithms (i.e., by deductive chains of statements).

Thinking one step larger, cities thus face emerging challenges: necessary city-related information is (mostly) described in natural language, which is often imprecise and uncertain. Sometimes there are no common definitions of certain terms. For efficient city development, it is crucial to include all urban information, but this task is not easy, as computer systems have difficulty computing such large data sets. It is increasingly important to enable 'creativity' in computer-based processes to help systems reason in a human-like way.

2.2 Given p and q, regarding inferential enchainment, only four cases can exist: $p<q$, $q<p$, both, and neither. Note that, as a statement, $p<q$ reflects the conditional statement 'If p, then q', and thus it can be said that, in $p<q$, p is the antecedent, or premise, or hypothesis, and q the consequent, or the consequence.

Correctly managing conditional statements is not easy—in childhood, for instance, it takes some time to correctly learn and manage them—nor are they always understood as being equivalent to an affirmative statement; it is also for this reason that it seems advisable to start with the 'negative' relation $\nless$ rather than with $<$, as we did.

When $p<q$, it is said that q follows forwards from p; when $q<p$, that q follows backwards from p; when both $p<q$ and $q<p$ hold, that p and q are deductively equivalent and written $p\approx q$; when neither $p<q$ nor $q<p$ hold, it is said that p and q are deductively isolated, and it is written $p \perp q$ or $q \perp p$.

In any case, if q is a conjecture from p, the following are the only possibilities:

- $p<q$, and q is an ordinary consequence of p;
- $q<p$, and q is a hypothesis for p, i.e., q 'explains' p; or
- $p \perp q$, and q is a speculation of p.

Hence, and independent of whether consequences are or are not particular conjectures, conjectures are perfectly classified into consequences, hypotheses, and speculations. In total, once a non-self-contradictory p is given, there are only refutations and conjectures of it. Thus, even maintaining doubt on the possible existence of consequences that are not conjectures, ordinary reasoning only consists of conjecturing and refuting, and conjecturing consists of deducing, abducing, and speculating, using the word abduction (cf. Peirce 1974) for the search for hypotheses.

2.3 Are hypotheses always conjectures? If $q < p$, then if $p < q'$ were to hold, provided the triplet (q, p, q') is transitive, it would follow that $q < q'$ and, provided q is not self-contradictory, it is concluded that $p \not\leq q'$. Hence, under transitivity, all non-self-contradictory hypotheses are conjectures.

Notice that consequences are obtained by forwards deduction from p ($p < q$) and hypotheses by backwards deduction from p ($q < p$). Concerning the existence of two contradictory hypotheses of p, that is, q and r, such that $q < p$, $r < p$, and $q < r'$, on one hand there is no 'formal' way (similar to the one followed with consequences) of falsifying it, and on the other hand anyone who has searched for an explanation for something knows that, often, contradictory hypotheses are possible. Notice that $q < p$, without $p < q$, seems to show that p 'contains' more than q, that a hypothesis is a simpler statement than what it explains; for this reason, neither p nor $q \approx p$ is taken as a hypothesis for p.

What about speculations from p? The characterization of $p \perp q$ does not allow examination of whether there are speculations that are not conjectures, but by limiting them to ones that are, it can be stated that, since then $p \not\leq q'$, there are only two types of such speculations:

- Speculations verifying $q' < p$ and $p \perp q$ (type-one speculations), and
- Speculations for which $q' \perp p$ and $p \perp q$ (type-two speculations).

Notice that in the first type of speculations, the negation q' can be reached by backwards deduction from p, and the problem lies in whether q can be deductively reached from q'.

(1) If $q < q'$, q is self-contradictory, and hence not interesting;
(2) If $q' < q$, it follows that $q' < (q')'$, and thus q' is self-contradictory and the process will stop;
(3) If $q \perp q'$, there is no deductive conclusion.

Notice also that the defining condition $q' < p$ implies $p' < (q')'$ and, if the negation is 'intuitionistic' $((q')' < q)$ or strong $((p)')' \approx p)$, and the triplet $(p', (q')', q)$ is transitive, then $p' < q$, and q can be deductively reached from p'. Thus, only under transitivity and intuitionistic or strong negation can a type-one speculation be deductively reached from p. For a type-two speculation, there is no deductive way to reaching it or its negation from p; type-two speculations always require jumps: they are 'deductively wild'.

In the main, speculations are reached not deductively but 'inductively'; finding them can be seen as creatively guessing, and speculations of type two are the most genuine creative instances, since neither they nor their negation has any deductive relation with p. They are deductively isolated from the starting information p furnishes. It is now clear that 'a jump' is understood to mean passing from p to a q such that p $\not\leq$ q, and q $\not\leq$ p, but p $\not\leq$ q'; that is, reaching a speculation that is neither a consequence nor a hypothesis but a conjecture (i.e., the third broken inequality). It should be noticed that, similar to hypotheses, there can exist pairs of contradictory speculations.

Finally, concerning refutations of q, since p < q', its negation is reachable from p by forwards deduction, and there is nothing against the existence of pairs of contradictory refutations. To check that a given q is a refutation, it suffices to recognize that p < q', but the actual problem is obtaining a refutation of p, like finding a consequence or a speculation or, in general, a conjecture rather than its checking.

All this reasoning requires both knowledge on the current subject and creativity; it is a serious problem, in which the point that is not debatable is the necessity of counting on some density of knowledge regarding p to understanding the refutations, consequences, and hypotheses of p. Reasoning without this neighborhood of knowledge regarding p is like trying to reach something from nothing. Experience on what surrounds p and previous experience of general searching are both important, and from this need is derived the relevance of learning and acquiring the best information possible on a subject before trying to research it.

George Siemens defined the concept of connectivism, in which he states that a single individual cannot reason without already having a knowledge base consisting of experiences and perceptions and that the reasoning process would perform even better if the experiences and perceptions of others could be included (Siemens 2005). Thus, the sum of all individual intelligences would help to extend the common knowledge base and facilitate reaching refutations or conjectures.

There is a lack of knowledge on what occurs when non-<-transitive triplets exist, as well as how to recognize it; that is, if in such a case the conjectures involved with p are or are not perfectly classified into ordinary consequences, hypotheses, and speculations. How can this problem be clarified? Possibly, it should be analyzed by means of controlled experiments in the pair (plain language, ordinary reasoning), something waiting to be obtained by means of true examples in ordinary reasoning and without imposing on it a mathematical model loaded with debatable laws. In summary, a scientific experiment-like study of creativity is currently open.

Let us pause for a while in the extreme cases. It can be supposed that the pair language-reasoning is endowed with a Boolean algebra or a De Morgan algebra structure (Trillas and García-Honrado 2013). The first model corresponds to the classical precise reasoning and the second to some type of reasoning with imprecise terms.

2.4 In the Boolean case, and as usual in classical logic, 'if p, then q' (p<q) is equivalent to p'+q = 1 ('not p or q' is a tautology), which is equivalent to p ≤ q, the natural lattice's order of the algebra given by p · q = p or, equivalently, by p+q = q. Since ≤ is a partial order, it is transitive, and in this case, transitivity is not local but universal and cannot conduct to linguistic continua; the Boolean model is unable to fully handle ordinary reasoning.

Such identification of conditionals is strictly linked with the laws generating the algebraic structure by the conjunction (·), the disjunction (+), the negation ('), the maximum (1), and the minimum (0). The Boolean algebra's structure corresponds to those lattices with negation that are loaded with more laws; hence, it is a very particular case appearing basically in the restrictive language of the mathematical proof but not always in the language of the mathematician's creative thought in which conjectures (and jumps) are usually made; nevertheless, jumps are not allowed in proving, even if some shortcuts are tolerated.

In the setting of Trillas and García-Honrado (2013), $p \leq q'$ coincides with $p \cdot q = 0$, and $p \leq q$ definitely implies $p \not\leq q'$, since if it were true that $p \leq q'$, it would follow that $p \cdot p = p \leq q \cdot q' = 0$, and thus $p = 0$, which is equivalent to $p \leq p'$; consequences are a strictly particular case of conjectures. In Boolean algebras, there is equivalence between contradiction and incompatibility. The only self-contradictory element is the minimum 0, and the only p whose negation is self-contradictory, $p' \leq (p')' = p'' = p$, is the maximum 1. Notice that $p \cdot q = 0$ is equivalent, with both $p \leq q'$ and $p \cdot q \leq (p \cdot q)'$.

The equivalence of contradiction and incompatibility is a serious difference from language where the concepts are independent. For instance, in the language associated with the logic of quantum physics in ortho-modular lattices (Bodiou 1964), contradiction implies incompatibility but not reciprocally; there are incompatible pairs that are not contradictory. Additionally, in fuzzy logic, the two concepts are independent, except in some particular fuzzy algebras in which either one implies the other, or the two are even equivalent (Trillas and Eciolaza 2015; Trillas 2017).

Analogously, non-null hypotheses are always conjectures. In fact, if $q \leq p$, then supposing $p \leq q'$, it would follow by transitivity that $q \leq q'$, equivalent to $q = 0$; thus, it must be true that $p \not\leq q'$. Notice that $q \leq p \leq q$ means $p = q$, and thus the only hypothesis that can be seen as a consequence is p; the intersection of the sets of consequences and hypotheses of p is the singleton $\{p\}$.

Concerning speculations, $p \perp q$ (or $p \perp q'$) simply means that p and q (or p and q') are not comparable under the lattice ordering $\leq$ of the Boolean algebra, and $q' \leq p$ is equivalent to $p' \leq (q')' = q$, that is, type-one speculations can be reached deductively forwards from not-p, the negation of the premise. Instead, type-two speculations can be obtained deductively neither from p nor from not-p. Type-two speculations are the ones that are properly inductive and require jumping to be obtained; under the Boolean algebra structure, creative reasoning is limited to obtaining type-two speculations.

2.5 In the case of De Morgan algebras, it is not valid to interpret 'If p, then q' by $p' + q$, since as soon as it is submitted to verify the Modus Ponens inequality $p \cdot (p' + q) \leq q$, from simply taking $q = 0$ it follows that $p \cdot p' = 0$ for all p, which only holds for the Boolean elements in the De Morgan algebra, hence forcing it to be a Boolean algebra. Taking $<$ as the lattice order relation $\leq$ of the De Morgan algebra, then if $p \leq q$, and thus $q' \leq p'$, provided $p \leq q'$, it would follow that $p \leq p'$, showing p to be self-contradictory but without implying $p = 0$, since in these algebras it is true neither that contradiction implies incompatibility nor that incompatibility always implies contradiction; the two concepts are independent in general. Anyway, by excluding self-contradictions, it is obtained that consequences are particular cases of conjectures. Analogously, if $q \leq p$, and supposing $p \leq q'$, it would follow that $q \leq q'$: that is, non-self-contradictory hypotheses are a particular case of conjectures, and, obviously, the only hypothesis that can be viewed as a consequence is p. Concerning type-one speculations, $q' \leq p$ implies $p' \leq (q')' = q$, showing that these speculations can be reached by forwards deduction from not-p, and hence that the only inductive speculations are of type two, the creative speculations.

2.6 Let us show again what follows (Trillas 2015):

a. Speculations make it possible to reach consequences and hypotheses by simply supposing the two laws $p \cdot q < p$, and $p < p + q$, that is, p always follows from 'p and q', and 'p or q' always follows from p; neither is rare at all.

b. In fact, if s is a speculation of p: $p \cdot s < p$, thus $p \cdot s$ is a hypothesis of p; and $p < p + s$, thus $p + s$ is a consequence of p. The conjunction of s with p gives a hypothesis, and its disjunction gives a consequence. Of course, it says nothing regarding the 'quality' of the hypotheses and the consequence, which depends on the corresponding 'quality' of the speculation.

c. If $h < p$, it is obvious that a consequence of p $(p < r)$ is also a consequence of h $(h < r)$, supposing the transitivity of the triplet (p, h, r). Thus, under transitivity, to show that h is not a hypothesis of p (à la Popper, to 'falsify h' as a hypothesis Popper 1963) when it cannot be directly seen that $h < p$, it suffices to find a consequence of p that is not a consequence of h.

d. Analogously, if $h < p$, and q is a conjecture from h, $h \not\leq q'$, then q is also a conjecture from p: otherwise, $p < q'$, and provided the triplet (h, p, q') is transitive, it would follow that $h < q'$, which is absurd. Hence, it should be true that $p \not\leq q'$, and q is a conjecture of p. Thus, under transitivity, to falsify h, it suffices to find one of its conjectures that is not a conjecture of p.

The first type of falsification is used in the formal sciences, the second in the experimental sciences. What is not stated in the former results is that all hypotheses (consequences) are decomposable in the form $p \cdot s$ $(p+s)$; in general, there are hypotheses and consequences that are not decomposable into such forms.

2.7 A possible alternative for $<$-transitivity is quasi-transitivity: $<$ is said to be quasi-transitive provided that $a < b$ and $b < c$ imply $a \cdot b < c$. If $<$ can be defined in the lattice style '$a < b \Leftrightarrow a \cdot b = a$', then $a < b$ and $b < c$ imply $a \cdot b = b$, and $a \cdot b < c$; that is, $<$ is quasi-transitive. Thus, in all the cases in which $<$ 'orders' the statements, as in all the lattice models and, for instance, in ortholattices and De Morgan algebras (and in particular in orthomodular lattices and Boolean algebras), the order relation $\leq$ is not only transitive but also quasi-transitive. Nevertheless, with fuzzy sets endowed with the typical, and transitive, pointwise ordering of membership functions, there is only quasi-transitivity with the t-norm min, the only t-norm ordering fuzzy sets, but not for any other t-norm representing the conjunction 'and'. A practical example with quasi-transitivity but without transitivity is offered in 2.1. The search for weak conditions to replace $<$-transitivity in such a form that the proven properties for $<$ can remain is an open topic.

3 Creative Reasoning

3.1 The discovery of the benzene ring by the German chemist August Kekulé was reached after a fantastic dream of a snake biting its tail suggested to him the chemical structure of the benzene molecule. The intellectual process leading to such a discovery can be qualified as true creative reasoning in science, not done emptily but by analogy and with the help of the knowledge Kekulé did actually have regarding benzene. It was a very fertile discovering, since it later led to the establishment the important German industry of colorants, and exemplifies what was previously mentioned regarding the neighborhood of positive and negative knowledge strictly required for creating. Kekulé made a jump based on his former knowledge that led him to propose a chemical structure that explained what had previously remained unexplained and whose practical consequences were unexpected (Rocke 2010).

Depending on the subject of the discovery, previous knowledge can be practical, theoretical, or a weighted mixture of the two. There is neither proof without logic nor a new finding without intuition based on analogy and knowledge; finding and proving are essential for the progress of scientific knowledge, which almost always requires speculation to arrive at abducing and deducing. Often enough, the crossing of previous knowledge on different areas helps to innovate with a new concept; for instance, the so-called 'fuzzy entropies' introduced by Aldo de Luca and Settimo Termini (cf. de Luca and Termini (1972)), where an innovation later made it possible to see the meaning of the word 'fuzzy' as a quantity, appearing as a measurable part of the vague concept of 'vagueness'. This innovation was possible thanks to cross-fertilization of the knowledge of de Luca and Termini (1972) (who were working in the interdisciplinary field of cybernetics) based on both Zadeh's fuzzy sets (Zadeh 1965) and Shannon's probabilistic entropy (Rényi 1961).

Good speculation is an art requiring some practice and, like all arts, is learned with (and against) a preceptor or master. To achieve fertile speculations, a researcher should be well educated and maintain a personal strong passion for study. Essentially, s/he should have taken the strong decision of mainly serving a field, which means constantly turning in her/his brain for some research goal; nobody can be considered a good researcher without, from time to time, presenting something new. For this purpose, good researchers do 'intellectually fight' against the best ones in their field. For this reason, both the process leading to the Ph.D. thesis and a good postdoctoral period elsewhere are now considered highly important, especially provided the first is done under a relevant senior researcher and the second by joining a challenging research team. The high quality of the educational environment is of great importance in science, as it is in engineering, humanities, architecture, medicine, art, and so on; it is a necessary but not a sufficient condition for reaching scientific relevance as a researcher; it is something that perhaps only a few great geniuses can avoid. These last statements hold not only for scientific activities but also for any other activity, whether of an intellectual type or not; regardless, creativity cannot be confused with talent or with erudition (Brentano 2016). In addition, it is important to note that researchers as well as creative people in various fields, as mentioned above, are crucial for the city's development. New (creative) ideas are required to progress the living standards of the citizens and to transform a traditional city into a smart and subsequent cognitive city.

3.2 The 'law' p < p can be easily accepted for all non-self-contradictory p, but without < being transitive for the triplets involved in the corresponding reasoning, almost all that has been shown could fail. Hence, and since the transitivity of < cannot be universally supposed for all triplets, as previously shown, what has been said should be considered 'local', not 'total' and is only 'universal' provided the transitive law is always valid. Hence, in ordinary reasoning and due to the lack of universal transitivity of <, it is possible to suppose the existence of statements p that are isolated, not connected by < with other statements q in either the forwards form p < q or the backwards form q < p; when p ⊥ q, 'creating q from p' is linked to the possibility of guessing q by speculation, which is the highly creative operation. It is speculation that facilitates to transcending traditional views by surpassing routine, mechanical research; this leap is actually one of the signs of creativity.

Let us consider for a while in an important characteristic of conjectures and refutations, namely monotony (Trillas 2015, 2017): that is, what occurs when information grows by adding more statements, passing from p to p · r, without this last being self-contradictory.

- In the case of consequences, if p < q, and the triplet (p · r, p, r) is transitive, then from p < q, and p · r < p (assuming this law as a natural one), it follows that p · r < q. Hence, the consequences of p are also consequences of p · r. 'Consequencing', like formal deduction, is monotonic: more information cannot lead to fewer consequences.
- With regard to hypotheses, if q < p · r, assuming p · r < p, then it follows (also assuming the transitivity of the triplet (q, p · r, p)) that q < p. Hence, the hypotheses of p are anti-monotonic: more information cannot lead to more hypotheses.
- Speculations of type one are also anti-monotonic: From q' < p · r, with p · r < p, it follows that q' < p; with more information, no more of these speculations can exist.
- Finally, speculations of type two are neither monotonic nor anti-monotonic, as can be easily shown with very simple examples in Boolean algebras or with sets; they are properly non-monotonic, as it cannot be formally stated what will occur when information is increased.
- Conjectures are, by and large, and assuming the same laws, anti-monotonic: From p · r ≰ q', supposing p < q', it would follow that p · r < q' (assuming p · r < p), which is absurd.
- In turn, refutations are monotonic: From p < q' and p · r < p, it follows that p · r < q'.

The monotonic law is satisfied by consequences and refutations, the anti-monotonic by all the conjectures, hypotheses and type-one speculations; only type-two speculations are 'wild' in the sense of neither being monotonic nor anti-monotonic, a view under which it can be said that the finding of type-two speculations is what constitutes the greatest part of induction, or inductive reasoning.

Of course, the lack of transitivity can cause the failure of such properties, and it indicates the necessity of some level of formalization to construct symbolic mathematical models for some parts of ordinary reasoning, as in the settings of ortholattices, De Morgan algebras, and basic algebras of fuzzy sets (Trillas 2017). It seems very difficult to imagine a symbolic mathematical model for 'all' the possible pairs of language-reasoning.

3.3 To end this section, let us add a naïve, imaginary, possibly artistic, and very risky hypothesis on the amount of knowledge needed to, through creative reasoning, arrive at a statement indicating some new concept.

Provided the set of knowledge on p, including analogies, consequences, hypotheses, speculations of both types, and refutations of p, that could be endowed with the topological structure of a metric space, then suppose such set of knowledge can be identified with a neighborhood $N(p, \Delta)$ of p, with a radius Δ. Were it possible to identify this radius as proportional to a given number, for instance, the Golden Number $\phi = (1 + 51/2)/2$: $\Delta = k \cdot \phi$, then $k = 1$ will indicate that the radius is coincidental to the Golden Number, and $k = \Delta/\phi$ could represent some kind of density of knowledge around p, increasing as Δ increases with respect to ϕ.

This (imaginary) idea is presented to indicate that perhaps, when much more experimental knowledge on reasoning is well known and available for comparison, it might not be impossible to arrive at some, for instance, metric topological models with parameters that can be calculated for each particular reasoning and perhaps characterize it by a series of such values. To reach this goal, it seems, as in other branches of science and when technology permitted it, that non-blind observations followed by controlled experiments on ordinary reasoning, and some formal hypotheses on the subject, might be previously designed and actually performed in the future. Today, when we can make use of highly sensitive sensors, powerful computers, and the Internet, it seems actually possible to begin such a task in a way similar, to some extent, to the one followed for studying the several forms in which the conjunction 'and' appears to be used in ordinary language (Guadarrama, et al. 2015). In other words, the variety of modern information and communication technologies can enable us and today's cities to foster creative activities to advance and become more efficient, sustainable and resilient.

In addition to a good education in art, music, science, or any other subject, passionate hard work is essential to achieve some degree of creativity. Pablo Picasso's famous answer "*Yes, but working!*" to the question "*Master, do you wait for the Muses?*" must not be forgotten. Some goals can be reached through rules, even through formulae and computation, but creativity is not one of them.

The words of Franz Brentano (Brentano 2016):

> Each human being has peculiarities in its form of binding ideas; like each one has an own form of walking, and an own style of writing, also his/her thinking enjoys an own form of reasoning,

likewise cannot be forgotten. As Brentano said, experimentation on human reasoning, particularly creativity, will be burdened with special difficulties; in any case, searching for 'neurological invariants' seems to be relevant.

4 Discussion

4.1 By its nature, this article cannot be conclusive, and even less does it pretend to be 'dogmatic'; currently, its subject does not allow 'doctrines'. Everything is only provisional, consisting of a trial for presenting some preliminary and provisional clues on what reasoning and creative reasoning are, and it can be compared to the practice musicians play before the conductor directs them to start the concert. Nevertheless, and metaphorically, for playing a concert on creative reasoning, there is not yet a score, and nor is there an orchestra, nor a conductor; the subject is in an initial process of gestation. This article contains nothing more than some naïve and incomplete thoughts and comments their authors risk writing before the still (scientifically) almost unknown territory, the land of reasoning in which creative reasoning lives, becomes fully explored; it tries to show that some theoretic (and partial) 'mechanisms' for creativity can be found, with the aim of fostering the transformation of an ordinary city into a cognitive city.

Notice that even using the word thought, thinking cannot be confused with reasoning; reasoning is performed thanks to thinking and language. Thinking comes from a natural brain phenomenon produced based on neural connections and is larger than reasoning; its study corresponds to the natural science of neurobiology, and reasoning is only one of its known manifestations. Creativity is not shown by a reproductive type of reasoning but by a productive one. As manifested in childhood, humans are inborn with creative skills, but a creative person is not just a talented one; in mankind, creativity evolves in degrees, the highest of which is 'the Genius'.

4.2 It seems clear that the lack of universal transitivity of the relation $<$ is what produces the existence of linguistic continua in language, similar to Poincaré's physical continuum in the sensory physical world (Poincaré 2011): something not at all bizarre once one considers that people actually reason with jumps and shortcuts not always coverable by a deductive path. That is, if the jump is from p to q, there may be no chain of statements $q_1, q_2, ..., q_n$, such that $p < q_1, q_1 < q_2, ..., q_n < q$ and, independently, that $<$ is or is not, transitive; if it is not transitive, then it cannot be concluded that $p < q$, but only perhaps $p \leq /q'$, guessing q.

Notice that in a situation with or without transitivity, provided $p \perp q$, and hence neither $p < q$ nor $q < p$ is true, the possibility of deductive paths still remains open, forwards, backwards, or both, from p to q, but not implying $p < q$. In any case, such paths of mixed backwards and forwards deduction are called 'heuristics' for arriving at the speculation q, from the premise p through a combination of forwards and backwards deduction. The following are two examples in a finite Boolean algebra with four atoms a_1, a_2, a_3, a_4, where the relation $\leq$ is transitive, by taking $p = a_1 + a_3$, and $q = a_1 + a_4$. As, indeed, $p \perp q$, and since $q' = a_2 + a_4$, it is also true that $p \perp q'$, and q is a type-two speculation from p. However, since $a_1 < p$, and $a_1 < q$, by going backwards from p to a_1 and then forwards from a_1, the speculation q is reached. A second example involves the same p, and $q = a_2 + a_4$, with $p \perp q$, and $q' = a_1 + a_3 = p$; thus, q is a type-one speculation of p reachable backwards from p to q', finally by negating q', since $(q')' = q = p' = a_2 + a_3$.

Notice that what is crucial for the possibility of doing a heuristic path like the first is to know $p \cdot q$ (in the example, a_1), or at least a part of it that is contained in q, which, in a linguistic situation, means the necessity of knowing something about what is searched for but being a part of the premise, counting on some partial information not isolated from the starting premise of the conjecture.

Usually, reasoning does not begin from a single premise but from a set $\{p_1, ..., p_n\}$ that nevertheless, in some way, could be summarized in a single one p (its résumé). For instance, after a previous ordering of the premises p_i, and since neither the associative nor the commutative laws are assumed for the conjunction $(\cdot)$, the following can be taken: $p = p_1 \cdot (p_2 \cdot (p_3 \cdot (... \cdot (p_n))))$, where the parentheses can be avoided when the operation is associative and the ordering when it is commutative. In addition, and since it is easy to prove that $p < p_i$ holds for all $i = 1, 2, ..., n$, each premise is a consequence of the résumé p, as proving it is necessary to suppose the transitivity of $<$. For instance, from $p < p_2 \cdot (...)))$, and $p_2 \cdot (...) < p_2$, under transitivity, it follows that $p < p_2$.

4.3 Some comments on the link between creative reasoning and truth are still in order: that is, what can be stated relative to the actual reality of conclusions in ordinary reasoning and, in particular, of speculations, based on the validity of the premises and, hence, of its résumé.

Even viewing truth as something neither universal nor absolute, to know what is true is pragmatically important. Applying the predicate true to statements is not only common in language but is important for at least knowing its degree of accuracy to reality. For this purpose, it is first necessary to capture a representation, indeed, a mathematical model (Trillas 2017) for the meaning of the word T = true.

For this word, as for all words, it is first necessary to know, usually in an empirical form, when a statement p is more, or less, true than another statement q; that is, to be able to recognize in language 'p is less true than q', symbolically shortened to $p <_T q$, which implies recognizing 'p is more true than q' $(q <_T p)$ and 'p is equally true as q' $(p <_T q$, and $q <_T p$, shortened to $p =_T q)$.

It should be remarked that the relation $<_T$ is usually not total or lineal, and hence there can exist pairs of statements p and q that are not comparable under $<_T$ but isolated under this relation that represents the qualitative meaning of T in the set S of considered statements, represented by the graph $(S, <_T)$. Thus, a measure of the extent to which a statement is true, that is, verifies the property of being true, is a mapping t assigning to each statement p a number $t(p)$ in $[0, 1]$ such that

(1) If $p <_T q$, then $t(p) \leq t(q)$;
(2) If p is maximal under $<_T$, that is, there is no statement truer than p, then $t(p) = 1$; and
(3) If q is minimal, that is, there is no statement less true than q, then
(4) $t(q) = 0$.

Note that t preserves the measure of those statements that are equally true, since $p =_T q$, or $p <_T q$ and $q <_T p$, implies $t(p) \leq t(q)$ and $t(q) \leq t(p)$, or $t(p) = t(q)$. Even if p and q are not comparable under $<_T$, the values $t(p)$ and $t(q)$ are actually comparable under the total order $\leq$ of $[0, 1]$.

The measures t, giving quantitative meanings of T in S, are also called 'degrees of truth', and in addition, once a measure t is known, the new lineal relation $\leq_t$, defined by $p \leq_t q \Leftrightarrow t(p) \leq t(q)$, clearly contains and enlarges the old relation $<_T$. The triplet constituted by the set S, the relation $<_T$, and a measure t, $(S, <_T, m)$ is simply a quantity reflecting a full-meaning of T in S; this approach is a way to scientifically domesticate meaning, and unless $<_T$ collapses with $= T$ it cannot be freely identified with $(S, \leq_t, m)$, since, for instance, there could be non-maximal statements with truth degree one.

The first three axioms of a measure, above, are insufficient to specify one of them; for that purpose, more reliable information, knowledge, or some reasonable hypotheses on the use of T should be added, for instance, the shape of the curve defined by t (for instance, that it is linear, quadratic, or another shape). The only exception (provided that there are maximal and minimal statements under $<_T$) is when $<_T$ collapses into $=_T$, that is, if the only recognizable relation between statements is that they are or are not equally true: that the use of T in S is precise. In such a case, S is perfectly classified in the subset of statements that are equally true and its complement; then, all the statements in the first class do have measure 1, and the statements in the second class have measure 0; the measure is unique and defines (specifies, in the sub-language of set theory) the set $t^{-1}(1)$, the set of true elements. This example is a limit situation reduced to certain special cases where concepts are given by necessary and sufficient conditions, as for almost all mathematical concepts, with some few exceptions such as the concept of a 'round number' (Renedo and Sobrino 2007).

Provided a degree of truth is known that is 'consistent with deduction', that is, if $< \subseteq <_T$, then $p < q => p <_T q => t(p) \leq t(q)$, the degree of truth of a consequence cannot be smaller than the degree of truth of the premise. Concerning hypotheses, $q < p => t(q) \leq t(p)$, the degree of truth of a hypothesis cannot surpass that of the premise.

In the case of refutations $p < q' => t(p) \leq t(q')$, the degree of truth of the refutation's negation cannot be smaller than the degree of truth of the premise, and supposing, for instance, $t(r') = 1 - t(r)$, for any statement r, it would follow that $t(q) \leq 1 - t(p) = t(p')$: the degree of truth of a refutation cannot surpass that of the premise's negation. If p were true, $t(p) = 1$, $t(p') = 0$, and $t(q) = 0$.

In general, for conjectures almost nothing can be said, since $p </q'$ does not allow the deduction of any inequality for t, except if $q' < p$; namely, if q is a type-one speculation.

In relation to speculation q, since $p \perp q$, nothing can be concluded with respect to the degrees of truth of q, which will nonetheless verify $t(p) \leq t(q)$, or $t(q) \leq t(p)$. For type-one speculations, if $q' < p => t(q') \leq t(p)$ then, provided it satisfies the law $t(r') = 1 - t(r)$, then $t(p') \leq t(q)$: the degree of truth of a type-one speculation cannot be smaller than that of the premise's negation. If $t(p') = 1$, then also $t(q) = 1$; however, surprisingly, $t(q) = 0$ implies $t(p) = 1$.

Concerning type-two speculations, since $p \perp q$ and $p \perp q'$, nothing can be presumed regarding the relation between $t(q)$ and $t(p)$; a priori, it is equally possible that $t(q)$ is larger or smaller than $t(p)$. It is not possible to foresee the truth limits of q; from the truth perspective, type-two speculations are also 'wild'. It should be noted that they are not directly based on either the total information supplied by p or in the total information supplied by p'; in some sense, they do follow from some imagining of the informative context in which p is inscribed, and good imagination is commonly linked with creativity. In the words attributed to Albert Einstein,

Logic will get you from A to Z, imagination will get you everywhere,

and in the words of Sagan (1980),

Imagination will often carry us to worlds that never existed, but without it we go nowhere.

4.4 Finally, it remains to reflect on how type-two speculations could be effectively reached, which is related to some common ideas around what is understood by 'imagination' and which, in the end, and in simple words, means forming a mental image or idea of something that is not yet clear or real. A short example on what solving a problem by analogy means, can be illustrative.

For instance, to 'imagine' a hypothesis h for p, that is, to 'see' a solution s to equation $h = s \cdot p$, the problem lies in finding such s provided it cannot deductively follow from p or p' and, hence, should be searched otherwise. How can this be done? A typical way, in human reasoning, consists of remembering a previously solved problem analogous to the current one, but based on a premise p^* that can be supposed to be similar to p, in which a solution s^* (and type-two speculation from p^*) was obtained, and supposing that some s is similar to s^*. That is, after recognizing that the current problem on p is analogous to the former on p^*, in which a satisfactory hypothesis $h^* = p^* \cdot s^*$ was found with p^* analogous to p, then, searching for s analogous to s^*, it is taken that $h = p \cdot s$. Of course, study, memory, and capability of seeing the analogy between the two problems are essential for arriving at the solution s.

Analogy, a natural resource for comparing different situations, is very important in ordinary reasoning; from very early ages, people can see analogies between faces, environments, houses, and so on; in some sense, it can be said that analogy is an engine for imagination. It makes it possible to connect situations that appear different; in Steve Jobs' words,

Creativity is just connecting things.

Since a typical scheme of analogy is the 'qualitative proportion of qualitative ratios' p: q :: r : s, where p is to q as r is to s, analogy can be seen to lie at the core of rationality.

There are many types of analogy, but analogy referring to reasoning, or inferential analogy, can be described by a mapping A between two universes of statements S and S^*, such that $p < q$ implies $A(p) = p^* < A(q) = q^*$; if q is a consequence of p, then q^* is a consequence of p^*. Thus, if $q < p$, then also $q^* < p^*$, that is, if q is a hypothesis for p, q^* is a hypothesis for p^*. Obviously, if $p \approx q$, then also $A(p) \approx A(q)$; analogy preserves the inferential equivalence of statements. Under this view, analogy appears as a morphism of deduction.

Provided A verifies the law $A(x') < A(x)'$, for all x in S, then if $p < r'$, it follows that $A(p) < A(r') < A(r)'$; that is, under transitivity, the analogy of a refutation of p is also a refutation of A (p); the same holds if the law is the stronger A (x') $\approx$ A (x)'. With the negative law [x $\not\leq$ y' => A (x) $\not\leq$ A (y)'], if p $\not\leq$ q', it follows that A (p) $\not\leq$ A (q)', by which if q is a conjecture of p, A (q) is so of A (p).

Hence, it seems that among the mappings A preserving deduction and abduction, only some of them verifying specific laws could be able to preserve refutations and conjectures; a formal study of how to obtain conjectures and refutations by analogy remains an open problem.

It should be noted that numerical control is relevant in analogy, that is, provided it is possible to know to what degree p and A(p) are analogous. For instance, considering only the attribute 'spherical shape', it can be said that oranges and apples are analogous, whereas by considering more attributes such as color, smell, taste, and so on, they are actually very different fruits.

Accordingly, to analyze the analogy or similitude of two classes of objects, it is essential to know as many common attributes as possible, as well as some that are not shared. In any case, analogy helps creativity, and in the words of Bohm (2004), a creative person

> is always open to learning what is new, to perceive new differences and new similarities, leading to new orders and structures, rather than always tending to impose familiar orders and structures.

To build analogies, it is essential to have a knowledge base from which existing and thus understood information can be retrieved to be matched with the newly acquired information. This matching makes it possible to comprehend the new information and to store it. Especially in the context of cognitive cities and their corresponding urban systems, it is crucial that people who are living and/or working in a city as well as the urban computer systems can construct analogies based on a common knowledge base. Therefore, it is crucial that people share their experience and perceptions to build an urban intelligence from which all relevant information can be retrieved.

5   Conclusions and Outlook

Creativity, based on human thinking, experience, memory, analogy, and hard work, can be considered to be associated with natural language and ordinary reasoning but not with artificial languages and formal reasoning. There are certainly forms of ordinary reasoning expressed in natural language that help to reach new conclusions without deductive paths, clearly calling for the famous Archimedes' shout 'Eureka!', the so-called 'ah-ha phenomenon' that nobody can doubt deserves to be qualified as creative. While there are some refutations, consequences and hypotheses among conclusions that are eventually obtainable in a creative but indirect way, the linkage with the premises by deductive paths means that stating them cannot always be seen as something surprising; what is actually surprising, and even unexpected, is to reach a conjecture in the class of speculations (especially of type-two) that are not obtainable from the premises by a deductive path, provided, of course, that they exist in the current problem. For instance, in the random case of throwing a die, whose events are represented by subsets in the finite Boolean algebra with atoms in the set $X = \{1, 2, 3, 4, 5, 6\}$, the résumé, corresponding to the elemental events obtainable at each throw, contains only the unique consequence X and the unique refutation Ø; the rest are hypotheses and are without speculations. Betting can be done on hypotheses, but neither on the sure event X (either 1, or 2, ..., or 6 will be obtained), nor on the absurd event Ø (no event will be obtained). Notice that the events $\{1, 2, 3\}$ (at most obtaining three), and $\{4, 5, 6\}$ (at least obtaining six), are contradictory since their intersection is empty, which simply illustrates the former statement on the existence of contradictory hypotheses.

Anyway, to qualify a reasoning as creative, the surprise its conclusion produces is essential, and this surprise is only truly achieved in the case of speculations, mainly with speculations of type two. It seems that it is in speculation and in the lack of transitivity where a possibility for studying creative reasoning actually lies; that is, in making non-deductive connections and escaping from preconceptions for which, to some extent, non-transitivity is responsible. Creativity demands, in any area, a break from preconceptions, and good creative reasoning leads to very fertile answers, largely in other fields (Menger 1974).

A remaining open problem is to perform a deeper analysis on the (usual) situation, in the reasoning expressed in natural language, in which too logical laws cannot be universally supposed, and not all the triplets of the involved statements are <-transitive; that is, by considering the pairs $(S, <)$ as 'linguistic continua', in which case the previously obtained results for conjectures and refutations can fail. It seems that such study requires, previously and necessarily, the help of experimental techniques, especially designed for such goals; without them, only more or less abstract and theoretical considerations can be included, but it is very difficult to believe that the highly complex phenomena of natural language and ordinary reasoning could be deeply studied and even modeled by means of just 'purely abstract thinking', always starting from something taken as obvious. It is similar to imagining that thermodynamics were created in a purely symbolical form, but without contrasting it with real phenomena.

Further exploration of creative reasoning is required to advance cognitive cities. Several fields (e.g., healthcare, mobility and logistics, security D'Onofrio and Portmann 2017) could profit from creativity, especially in the development of new ways to make the city more resilient. More understanding of natural language and ordinary reasoning and, in particular, of creative reasoning as an important side of creativity is actually basic for reaching (urban) computer systems to 'think like people'; in summary, to undo the so-called Gordian Knot of Artificial Intelligence.

Is a new frontier for science in front of us? In the end, with the words of Bohm (2004),

> Just as the health of the body demands that we breathe properly, so, whether we like it or not, the health of the mind requires that we be creative.

To be a creative person, working like a craftsman is highly recommended; the words of Isaac Asimov,

As far as creativity is concerned, isolation is required,

cannot be forgotten (Asimov 2014). Group discussion and brainstorming can be interesting to booster creativity, but in the end, creation is an individual task whose teaching only seems possible through close contact with a very creative person. In the words attributed to Albert Einstein,

Creativity is contagious, pass it on.

Acknowledgements The authors would like to thank the anonymous reviewers whose comments helped to improve this paper.

References

Asimov I (2014) How do people get new ideas? MIT Technological Review
Bodiou G (1964) Théorie dialectique des probabilités. Gauthier-Villars, Paris
Bohm D (2004) On creativity. Routledge, London
Brentano F (2016) La genialidad, Ediciones Encuentro, Madrid (Translation into Spanish of the 1892's German 'Das Genie')
De Luca A, Termini S (1972) A definition of a nonprobabilistic entropy in the setting of fuzzy sets theory. Inf Control 20:301–312
D'Onofrio S, Portmann E (2015) Von fuzzy-sets zu Computing-with-words. Informatik-Spektrum 38:1–7
D'Onofrio S, Portmann E (2017) Cognitive computing in smart cities. Informatik Spektrum 40(1):46–57
Guadarrama S, Renedo E, Trillas E (2015) A first inquiry on semantic-based models of and. In: Seising R et al (eds) Accuracy and fuzziness. Springer, Switzerland, pp 331–350
Menger K (1974) Morality, decision and social organization. toward a logic of ethics. Reidel, Dordrecht
Moyser R, Uffer S (2016) From smart to cognitive: a roadmap for the adoption of technology in cities. In: Portmann E, Finger M (eds) Towards cognitive cities: advances in cognitive computing and its applications to the governance of large urban systems. Springer International Publishing, Heidelberg, pp 13–35
Peirce CS (1974) Collected papers of Charles Sanders Peirce. Harvard University Press
Poincaré H (2011) Science and hypothesis. Dover, New York
Popper KR (1963) Conjectures and refutations: the growth of scientific knowledge. Routledge & Kegan Paul, London
Portmann E, Finger M (2015) Smart cities–Ein Überblick! HMD Praxis der Wirtschaftsinformatik: 1–12
Renedo E, Sobrino A (2007) Round numbers revisited. A new fuzzy approach. Fuzzy Sets Syst 158:1618–1629
Rényi A (1961) On measures of entropy and information. In: Proceedings of the fourth Berkeley symposium on mathematical statistics and probability
Robinson K, Aronica L (2016) Escuelas creativas. Random House, Barcelona
Sagan C (1980) Cosmos. Random House, New York
Siemens G (2005) Connectivism: a learning theory for the digital age. Int J Instr Technol Distance Learn 2(1):3–10
Rocke AJ (2010) Image and reality: Kekulé Kopp, and the scientific imagination. University of Chicago Press, Chicago

Sternberg RJ (2003) Wisdom, intelligence, and creativity synthesized. Cambridge University Press, Cambridge

Tarski A (1956) Fundamental concepts of the methodology of deductive sciences, logic, semantics, metamathematics. Oxford University Press, Oxford

Trillas E (2015) Glimpsing at guessing. Fuzzy Sets Syst 281:32–43

Trillas E (2017) On the logos: a naïve view on ordinary reasoning and fuzzy logic. Springer, Berlin

Trillas E, Eciolaza L (2015) Fuzzy logic. Springer, Berlin

Trillas E, García-Honrado I (2013) ¿Hacia un replanteamiento del cálculo proposicional clásico? Agora 32(1):7–25

Wiles A (1995) Modular elliptic curves and Fermat's last theorem. Ann Math 141:443–551

Zadeh LA (1965) Fuzzy sets. Inf Control 8(3):338–353

Enric Trillas currently researches at the University of Oviedo, Spain. He was formerly Full Professor at the Technical Universities of Catalonia and Madrid as well as a researcher at the European Center of Soft Computing. He is the editor and author of a dozen books and has written more than 400 articles in journals, editorial books and conference proceedings. He is a member of the editorial board of several national and international journals. Enric Trillas is a fellow of IFSA and honorary member of EUSFLAT, a Distinguished Visiting Professor of the National University of Córdoba (Argentina), Doctor "Honoris Causa" of the Universities Pública de Navarra and Santiago de Compostela and member of the Accademia Nazionale di Scienze, Lettere e Arti of Palermo. He has won numerous national and international awards, including the IFSA's Outstanding Achievement Award, the Kampé de Feriet Medal, the Fuzzy Pioneer Award from EUSFLAT and IEEE-CIS. Enric Trillas served his country in various government positions: He was Spain's Delegate in the OECD's Committee for Scientific Research in Paris, President of the Spanish High Council for Scientific Research (CSIC), Director General of the National Institute for Aerospace Technology (INTA), and General Secretary of the National Plan of Research and Development. His current research is on the theoretical foundations of fuzzy logic, ordinary reasoning, and creativity. His last book is "On the Logos: A Naïve View on Fuzzy Logic and Ordinary Reasoning" (2017). His new book "El desafío de la creatividad" is in printing.

Sara D'Onofrio is PhD-student in Computer Science at the Human-IST Institute of the University of Fribourg, Switzerland, and IT trainee at Swiss Post. She has a bilingual bachelor's degree in Business Administration from the University of Fribourg and a master's degree in Business Administration with a specialization in Business Information Systems from the University of Bern, Switzerland. She has attended courses in business modeling (University of Vienna, Austria), fuzzy logic (Lake Como School of Advanced Studies, Italy; Centro Singular de Investigacion en Tecnoloxias da Informacion, Spain), fuzzy cognitive maps (Volos, Greece) and design science and design thinking (Ovronnaz, Switzerland). She took part in several conferences in Europe, South America and Canada. She has written more than 20 articles in journals, editorial books and conference proceedings and is student member of EUSFLAT and IEEE. Her current research interests lie in soft and cognitive computing, natural language processing, and smart and cognitive cities.

Edy Portmann is a researcher and scholar, specialist and consultant for semantic search, social media, and soft computing. Currently, he works as a Swiss Post-Funded Professor of Computer Science at the Human-IST Institute of the University of Fribourg, Switzerland. Edy Portmann studied for a BSc in Information Systems at the Lucerne University of Applied Sciences and Arts, for an MSc in Business and Economics at the University of Basel, and for a PhD in Computer Sciences at the University of Fribourg. He was a Visiting Research Scholar at National University of Singapore (NUS), Postdoctoral Researcher at University of California at Berkeley, USA, and Assistant Professor at the University of Bern. Next to his studies, Edy Portmann worked several years in a number of organizations in study-related disciplines. Among others, he worked as Supervisor at Link Market Research Institute, as Contract Manager for Swisscom Mobile, as Business Analyst for PwC, as IT Auditor at Ernst & Young and, in addition to his doctoral studies, as Researcher at the Lucerne University of Applied Science and Arts. Edy Portmann is repeated nominee for Marquis Who's Who, selected member of the Heidelberg Laureate Forum, co-founder of Mediamatics, and co-editor of the Springer Series 'Fuzzy Management Methods', as well as author of several popular books in his field. He lives happily married in Bern and has three lively kids.

Using Fuzzy Cognitive Maps to Arouse Learning Processes in Cities

Sara D'Onofrio, Elpiniki Papageorgiou and Edy Portmann

Abstract Processing information in a city is simultaneously a primary task and a pivotal challenge. Urban data are usually expressed in natural language and thus imprecise but can contain relevant information that should be processed to advance the city. Fuzzy cognitive maps (FCMs) can be used to model interconnected and imprecise urban data and are therefore suitable to both address this challenge and to fulfil the primary task. Cognitive cities are based on connectivism, which assumes that knowledge is built through the experiences and perceptions of different people. Hence, the design of a cognitive learning process in a city is crucial. In this article, the current state-of-the-art research in the field of FCMs and FCMs combined with learning algorithms is presented based on an extensive literature review and grounded theory. In total, 59 research papers were gathered and analyzed. The results show that the application of FCMs already facilitates the acquisition and representation of urban data and, thus, helps to make a city smarter. However, using FCMs combined with learning algorithms optimizes this smartness and helps to foster the development of cognitive cities.

Keywords Cognitive city · Connectivism · Fuzzy logic · Fuzzy cognitive maps
Learning algorithms · Smart city

1 Introduction

Cities face the challenge of big data, a term that denotes data sets that, due to their size, cannot be processed and analyzed through conventional hardware (Chen et al.

S. D'Onofrio (✉) · E. Portmann
Human-IST Institute, University of Fribourg, Boulevard de Pérolles 90, 1700 Fribourg,
Switzerland
e-mail: sara.donofrio@unifr.ch

E. Papageorgiou
Department of Electrical Engineering, University of Applied Sciences (TEI) of Thessaly, Larisa,
Greece

© Springer Nature Switzerland AG 2019   107
E. Portmann et al. (eds.), *Designing Cognitive Cities*, Studies in Systems,
Decision and Control 176, https://doi.org/10.1007/978-3-030-00317-3_5

2012). Thus, the amount of available data in a city is so large that it becomes difficult to use the collected information in a way that increases its utility. Therefore, methods to enhance data quality, efficient filters and intelligent algorithms are required to acquire the relevant information. Additionally, existing information and communications technologies (ICTs) (e.g., mobile applications and question-answering systems) have to be improved to address the challenge of big data. In this way, cities are enabled to analyze, use and save the information that they collect from their citizens through advanced ICTs. It is even more important that cities learn from this information to be able to understand and reflect on their urban processes. The knowledge that humans and systems possess can be broadened when they share it and learn from each other. This continuous information exchange enables the building of an urban intelligence (i.e., a collective intelligence in a city) using the approach of connectivism (i.e., learning and cognition theory), assuming that knowledge increases through experiences and perceptions and enables actors to act as a unit (i.e., a cooperating group) (Malone and Bernstein 2015; Siemens 2005).

Much of world knowledge is expressed in natural language and is thus ambiguous. In other words, the city is not only confronted with large amounts of data but also with their ambiguity. People generally use words or sentences to describe their experiences (i.e., things they have seen, heard, smelt and felt), which can be described as perceptions (Zadeh 2006). Fundamentally, perceptions are imprecise, and thus, the management of those perceptions is a challenging task. To cope with this, it is advisable to use fuzzy cognitive maps (FCMs) (Kosko 1986).

FCMs consist of nodes which represent concepts or objects and edges which represent fuzzy relationships between concepts or objects, which are aggregated into a graph-like structure. By allowing for the addition of attributes to nodes and edges, as well as by providing a (modifiable) feedback through the display of fuzzy-weighted digraphs, FCMs make it possible to model complex urban constructs (D'Onofrio et al. 2017; Kosko 1986). This can be very important to the design and development of cognitive cities because, following the law of parsimony (Laird 1919), FCMs offer a way to model real-world concepts.

Cognitive cities are seen as an advancement of smart cities. A cognitive city uses, similar to a smart city (cf. Portmann and Finger 2015), advanced ICT to collect unstructured, semi-structured and structured urban data to analyze, understand and apply them to improve its urban processes based on available data (i.e., considering citizens' needs). The primary difference between a smart and a cognitive city is an additional component to the cognitive city. By adding cognition (i.e., cognitive processes) to a city, the city can not only collect, analyze and re-use its data for its own citizens but also learn and reflect on the collected urban data (D'Onofrio et al. 2017).

However, these learning experiences can only be successful if other information, along with human experiences and perception, is also considered (Siemens 2005). Thus, it is crucial to build connections (i.e., networks) between humans as well as between humans and computer systems (e.g., by means of FCMs [Kosko 1986]) (D'Onofrio et al. 2016). By building those connections, for example, through intelligent (i.e., cognitive) systems, a city can steadily learn and improve itself. Learning algorithms are, in this context, of great interest. By adding such algorithms, FCMs can be enhanced and, through the provided feedback loop, complemented with learning processes. Increasing numbers of cognitive systems, such as IBM Watson (Kelly III 2015) or the personal digital assistant 2.0 (Kaltenrieder et al. 2016), have been developed to improve the interaction between cities and their citizens. Therefore, it is essential that they can learn from each other and that this new knowledge can be stored in an efficient way (D'Onofrio et al. 2017).

In the style of Papageorgiou and Salmeron (2013), the primary aim of this article is to give a basic understanding of FCMs and learning algorithms and a review of FCM applications, with or without learning algorithms, in the field of smart and cognitive cities. The intention of the authors is to improve the data-driven processes of cities. Therefore, only FCM applications within cities will be covered in this study.

This article is structured as follows. Section 2 presents the theoretical background by introducing soft computing, learning algorithms and cognitive cities, followed by the description of the methodology in Sect. 3. In Sect. 4, a selection of existing smart applications using FCMs is presented, and Sect. 5 provides an overview of applications using FCMs and learning algorithms. Section 6 illustrates how learning algorithms help to foster the development of a city into a cognitive city. Section 7 concludes the article and provides insights into the limitations of the concepts and suggestions for further research.

2 Theoretical Background

To provide a base of knowledge for the following analysis, this section offers insight into the theoretical concepts of soft computing, learning algorithms and cognitive cities to reach a better understanding of this article.

2.1 Components of Soft Computing

This section presents important concepts of soft computing, including fuzzy logic and fuzzy sets, FCMs, granular computing and computing with words. The primary focus is on FCMs. The concepts of fuzzy logic and fuzzy sets serve as an introduction, whereas the use of granular computing and computing with words can be seen as a possible extension of FCMs.

2.1.1 Fuzzy Logic and Fuzzy Sets

Because humans are confronted with imprecise, uncertain and vague information (as world knowledge is mostly described in natural language), a concept of logic is required that is able to meet this challenge. The traditional bivalent logic, consisting of two truth values (i.e., *true* and *false*), is not able to accomplish this task. To tackle this shortcoming, Zadeh (1965) introduced fuzzy set theory, the foundation of the concept of fuzzy logic (Zadeh 1988).

In traditional set theory, a set is defined by its elements (i.e., either it belongs to a set or it does not). Unlike the classical set, a fuzzy set is not described through its elements. This implies that every element is in every fuzzy set, but only to a degree. Thus, via calculation with degrees, an element has different degrees of membership in several fuzzy sets (D'Onofrio and Portmann 2015; Zadeh 1965).

A fuzzy set is determined by the membership function $f(x)$, which maps an arbitrary element x to the real interval [0,1] (i.e., values between 0 and 1) (Zadeh 1965). The closer the value of the element is to 1, the higher its degree of membership to the fuzzy set; the closer the value is to 0, the lower its degree of membership to the fuzzy set. To converge this concept to the real world, linguistic variables can be used as labels for the membership degree (e.g., *low, little,* and *close*). A linguistic variable is hereby considered fuzzier since it can represent an uncertain value (D'Onofrio et al. 2017). Therefore, fuzzy logic augments bivalent truth by partial truth (i.e., allowing an element to be anything between *true* and *false*) (Zadeh 1988).

2.1.2 Fuzzy Cognitive Maps

Built on fuzzy logic, FCMs is another method of soft computing. FCMs emerged out of (traditional) cognitive maps (cf. Tolman 1948), a concept describing a set of psychological transformations by which a human handles information and its attributes in her/his everyday environment (Downs and Stea 1973). This mental representation can be used as a modeling tool in which nodes C_i embody concepts or objects, and their relationships m_i are represented by directed edges in a graph structure, which are generally designed as multigraphs (i.e., one node can have multiple incoming edges) (D'Onofrio et al. 2017; Kosko 1986).

Enriched with fuzzy logic (Zadeh 1988), nodes C_i and edges m_i are allowed to have fuzzy states (e.g., *bad weather, high risk* or *low, little, close* instead of numbers). For this, the weights that contain fuzzy values (i.e., $\in [-1,1]$) are assigned to the edges and describe (with numerical or linguistic variables) the relationship between nodes C_i:

$$g(x) = \begin{cases} positive\,relationship, & x > 0 \\ no\,relationship, & x = 0 \\ negative\,relationship, & x < 0 \end{cases} \tag{1}$$

In turn, these edges m_i between the nodes C_i can be represented as an adjacency matrix, showing their degree of causal relationship. For example, the element m_{kl} in matrix M denotes the weights of the edge m_{kl} that connects the two nodes C_k and C_l and takes values between -1 and 1, denoting a negative or positive effect of C_k on

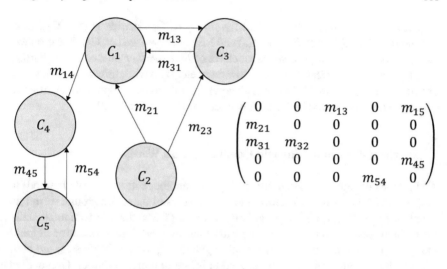

Fig. 1 General example FCM with its adjacency matrix

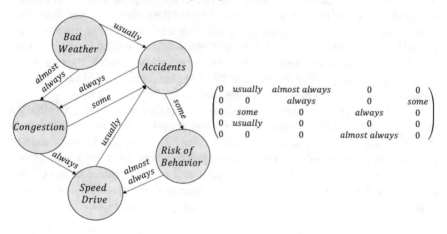

Fig. 2 Practical example FCM with its adjacency matrix

C_l. If there is no effect, the element is equal to zero (Kosko 1986). Figure 1 shows a general example of an FCM and its adjacency matrix, from Kaltenrieder et al. (2016).

As FCMs influence each other, they change over time (i.e., are dynamic models). Because of their dynamic adaptation (e.g., using generic rules, crisp relations or fuzzy rules [Aguilar 2013]), FCMs are able to simulate complex real world constructs (Kannappan et al. 2011). Furthermore, the use of fuzziness helps to model imprecision in a way that classical logic cannot (Zadeh 1988). Figure 2 illustrates another FCM, this time, with fuzzy states from Aguilar (2013) to substantiate the advantage of fuzziness.

The concepts of this FCM are *Bad Weather* (i.e., C_1), *Accidents* (i.e., C_2), *Congestion* (i.e., C_3), *Speed Drive* (i.e., C_4) and *Risk of Behavior* (i.e., C_5). For instance, the edge *usually* (i.e., m_{12}) denotes an effect of *Bad Weather* on *Accidents*. Further *Accidents* has *some* effect on *Risk of Behavior*, etc. By having linguistic values, $P = \{some \leq usually \leq almost\ always \leq always\}$, the effects (i.e., edges) of the different concepts (i.e., nodes) can be intuitively understood (Aguilar 2013).

2.1.3 Granular Computing and Computing with Words

Adding the concept of granular computing (i.e., another soft computing method) to FCMs enables humans to transform the complexities of the world around them into simple theories (i.e., different granularity levels in FCMs) that are computationally tractable to reason (Hobbs 1985). By partitioning an object into a collection of granules (i.e., a clump of objects), information granules (possibly fuzzy sets) are built (Zadeh 1998), which, in turn, can be used in FCMs to represent nodes. Granulation helps to merge incoming data, which is an important component of FCMs because the inputs (mostly of experts) are a crucial aspect in the design of FCMs. Therefore, it is important to take into account the granulation process as well as the possibility of building different granularity levels in FCMs to allow zooming in on (i.e., more details) and zooming out from (i.e., less details) activities to mimic the real world. The ability to conceptualize the world at different granularities and to switch among them is fundamental for human reasoning and flexibility (Hobbs 1985).

Granular computing serves as a basis for computing with words, a technique that is tolerant regarding imprecision and that allows us to compute with words and sentences instead of numbers (Zadeh 1996). This implies the translation of propositions from natural language into computable generalized constraints (Zadeh 2008). Thinking a step further, this computational method enables us to consider all available (urban) data, even if they are characterized by imprecision, uncertainty, vagueness and partial truth, which can then be integrated into FCMs.

2.2 Learning Algorithms

There exists no generally accepted definition of *learning*. A learning process entails the acquisition of new declarative knowledge, which is then tested, converted, stored for future use, and applied to make changes reflecting the emotional, social and societal aspects within the given environment (Campbell 2009). Since the beginning of the computer era, it has been the aim of researchers to develop computer systems that are able to learn in a human-like way. Learning algorithms might have a great potential to enable learning processes in FCMs and can be seen as a first step in the direction of cognitive systems.

The determination of the weight matrix (i.e., the relationships among concepts in a FCM) of a given FCM is crucial because the robustness and accuracy of information is highly dependent on it. Learning algorithms have become an important component of FCMs because they allow the adjustment of the operation and accuracy in a number of modeling and prediction tasks (Papageorgiou 2012). Therefore, learning algorithms can be used to adjust weights in FCMs to correctly predict or classify unknown examples and, thus, make it possible to work with the real world.

Learning algorithms can be supervised (e.g., ensemble methods, naïve Bayes classification, and support vector machines), reinforcement (e.g., learning automata, q-learning, and temporal difference learning) or unsupervised (e.g., clustering algorithms, expectation-maximization algorithms, and Hebbian learning) (Papageorgiou 2012). Some learning algorithms are already applied to FCMs (cf. Sect. 5) and help to improve existing FCMs by adjusting their weight matrix to update the initial knowledge of human experts, to change them according to the specific problem characteristics and/or to take into consideration any knowledge from historical data to build learned weights (Kannappan et al. 2011; Papageorgiou 2012).

2.3 Cognitive Cities

Mostashari et al. (2011) first used the term cognitive city in relation to smart cities and e-governance. Hereby, cognitive cities are seen as an advancement of smart cities by introducing a new component, *cognition,* in the form of cognition and learning theories and methods (e.g., connectivism [Siemens 2005]) to the constant development of urban processes. Thus, leveraging ICT and imbedded intelligence, coupled with innovative service provision and governance structure, can allow cities to cope with complex challenges, such as big data (and fuzziness in information), urbanization and digitalization. This means that cognitive cities rely on cognitive processes and systems that can learn from past experiences and respond adequately to changes in their environments (Hurwitz et al. 2015). To enable such a learning process, it is important that the stakeholders of the city (e.g., citizens, out-of-town visitors) and their interactions (e.g., with other citizens as well as with computer systems) are the center of focus (D'Onofrio et al. 2017; Kaltenrieder et al. 2015).

By applying the learning and cognition theory *connectivism* (Siemens 2005), citizens not only learn based on their own experiences and perceptions but also have to rely on the experiences and perceptions of others. Thus, in a cognitive city, the citizen becomes an active element through civic participation as well as through using urban infrastructure and processes (e.g., citizen as a sensor [Goodchild 2007]). Sharing these data enables the city to understand, reflect, analyze and optimize existing urban processes based on the citizens' needs. In this case, the component, *cognition,* enables the existence of learning, memory creation and experience retrieval to continuously improve urban processes and to thus foster the development of cognitive cities (D'Onofrio et al. 2017; Mostashari et al. 2011).

3 Methodology

First, a literature review was performed, which is defined as an evaluative report of prior, relevant literature related to one's own research area (Webster and Watson 2002). The literature search was performed online using various science platforms, such as Google Scholar,[1] Elsevier Science Direct,[2] and Association for Computing Machinery[3] and search engines such as Google.[4] Using this method, a selection of approximately 150 papers were gathered. After examining them, 59 papers were selected based on following criteria:

(1) FCMs as an applied technique (must)
(2) City as a referred field (must)
(3) Learning algorithms used to complement FCMs (preferred)

The aim of this analysis is to depict the current state-of-the-art research on FCMs and FCMs combined with learning algorithms in the field of smart and cognitive cities. It should be emphasized that it is pivotal to first evaluate the current state and to then build on it. The reason for this procedure is that the existing research data contain valuable information that can be reused.

Second, the selected papers were examined to see why and for which purpose FCMs and learning algorithms were used. For this, grounded theory was applied: a methodology for developing a theory based on data that are gathered and analyzed systematically (Strauss and Corbin 1994). Therefore, every study was analyzed in more detail to understand how these techniques might address city-related issues. This implies using the advantages of both techniques (i.e., FCMs and learning algorithms) and thus deriving suggestions regarding how the combination of FCMs and learning algorithms can advance cognitive cities.

4 Smart Applications Using Fuzzy Cognitive Maps

This section provides an overview of existing smart applications using FCMs to enhance urban processes. For ease of comprehension, the applications are categorized into convenient fields, summarized in Table 1,[5] and are shortly described for the final analysis in Sect. 6. A total of 40 research studies (not including the FCM applications using learning algorithms) have been selected, which have been categorized into five fields—business (19), ecology (7), education (2), intelligent systems (5) and medicine (7)—to be presented in this chapter.

---

[1]Cf. https://scholar.google.ch/.

[2]Cf. http://www.sciencedirect.com/.

[3]Cf. http://dl.acm.org/.

[4]Cf. https://www.google.ch/.

[5]Please note: the left column always states the source and the right the applied field.

Table 1 FCM applications

Business	
Altay and Kayakutlu (2011)	Factor priorization
Chytas et al. (2011)	Performance measurement
Çoban and Seçme (2005)	Strategic planning
Hanafizadeh and Aliehyaei (2011)	Expert system
Irani et al. (2002)	Investment justification
Jetter and Schweinfort (2011)	Scenario planning
Kang et al. (2004)	Relationship management
Kardaras and Karakostas (1999)	Strategic planning
Lazzerini and Mkrtchyan (2010)	Risk analysis
Lee et al. (2004)	Performance measurement
Rodriguez-Repiso et al. (2007)	IT project evaluation
Salmeron (2009)	Project selection
Salmeron and Lopez (2012)	Risk management
Sharif and Irani (2006)	IS evaluation
van Vliet et al. (2010)	Scenario planning
Xirogiannis and Glykas (2007)	E-business
Xirogiannis et al. (2004)	Urban design
Yaman and Polat (2009)	Effect-based planning
Yu and Tzeng (2006)	Decision-making
Education	
Georgiou and Botsios (2008)	Online learning
Hossain and Brooks (2008)	Software adaption
Ecology	
Banini and Bearman (1998)	Slurry rheology
Bertolini (2013)	Industrial ergonomics
Kok (2009)	Forest management
Kok et al. (2000)	Water management
Kottas et al. (2006)	Water management
Skov and Svenning (2003)	Forest management
Tan and Özesmi (2006)	Environmental management
Intelligent Systems	
Acampora and Loia (2009)	Computational intelligence
Lee et al. (2002)	Web-mining technology
Lo Storto (2010)	Software development
Parenthoen et al. (2001)	Interactive virtual worlds
Peláez and Bowles (1996)	Complex system

(continued)

Table 1 (continued)

Medicine	
Georgopoulos et al. (2003)	Language impairment, dyslexia, autism
Giles et al. (2007)	Diabetes
Iakovidis and Papageorgiou (2011)	Expert system
Innocent and John (2004)	Expert system
Papageorgiou et al. (2009)	Infectious diseases
Rodin et al. (2009)	Bioinformatics
Stylios and Georgopoulos (2008)	Language/speech disorders, obstetrics

4.1 Business

Altay and Kayakutlu (2011) chose FCMs as a method to reduce factors or criteria (i.e., factor prioritization) in a decision-making environment to counteract the challenge of making decisions with too many factors. In this case, FCMs help to establish a realistic and robust decision making process. Yaman and Polat (2009) used FCMs to support the decision-making process in effect-based planning (i.e., new force application concept in military planning), and Yu and Tzeng (2006) applied FCMs to multi-criteria decision-making with dependence and feedback effects to overcome the shortcomings of analytic network processing (Saaty 1996). Thus, they added FCMs to an existing network-like method for optimization reasons. To model specialist knowledge to contribute to the effort to develop more intelligent control methods and autonomous decision support mechanisms in urban areas, Xirogiannis et al. (2004) also employed FCMs. In doing so, the process of esthetic analysis in urban areas was improved.

Sharif and Irani (2006), however, applied FCMs to the modeling of tangible and intangible aspects of the information system evaluation decision-making task to elucidate key aspects of an investment justification process. In the study of Salmeron (2009), he proposed the use of FCMs to make predictive comparisons among research projects under evaluation in an attempt to support the project selection. In the study of Lee et al. (2004), they stated that FCMs could be used to describe the inference process for evaluating electronic data interchange performance. Within the topic, Rodriguez-Repiso et al. (2007) identified, classified and evaluated critical success factors of IT projects by implementing FCMs. Hanafizadeh and Aliehyaei (2011) applied FCMs to combine the views of experts and to aggregate these FCMs into one FCM to offer recommendations and changes in soft systems methodology.

FCMs can be used not only for decision-making, evaluation or expert systems but also for strategic planning tasks. For instance, Çoban and Seçme (2005) used FCMs to predict the socioeconomic consequences of privatization at the firm level, while Jetter and Schweinfort (2011) applied FCMs to facilitate scenario planning by integrating the qualitative and partial knowledge of multiple individuals. Kardaras and Karakostas (1999) implemented FCMs to comprehensively model primary knowledge and to simulate and evaluate several alternative ways to use IT to improve organizational performance. However, Van Vliet et al. (2010) employed FCMs as a semi-quantitative communication and learning tool in scenario studies to structure participatory output, which provides a base for quantification.

In the area of finance, focusing on investments, Irani et al. (2002) applied FCMs to model each IT/IS evaluation factor, integrating strategic, tactical, operational and investment factors, to highlight interdependencies among contributory justification factors. However, Lazzerini and Mkrtchyan (2010), as well as Salmeron and Lopez (2012), used FCMs to analyze the relationship between risk factors and risks to forecast possible risk effects.

Finally, Chytas et al. (2011) used FCMs for a proactive balanced scorecard (i.e., evaluating the impact of each key performance indicator), Kang et al. (2004) implemented FCMs to describe the inference process for the management of relationships among organizational members in an airline service, and Xirogiannis and Glykas (2007) employed FCMs as the underlying methodology to quantify the impact of strategic changes on the overall e-business efficiency.

4.2 Ecology

Skov and Svenning (2003) applied FCMs in combination with geographic information system-based spatial operations to produce gradient maps based on standard forestry maps and expert knowledge in forest management to predict ground flora species richness under various forest management scenarios. Another approach in forest management was proposed by Kok (2009), using FCMs to capture (future) dynamics of deforestation by integrating key factors and feedbacks into the model. Such a model can also be found in the research study from Bertolini (2007). In his study, he used FCMs to assess the most important factors for operators in the management and control of production plants.

Tan and Özesmi (2006) used FCMs to capture the behavior of a shallow lake ecosystem as a whole. This was achieved by gathering the opinions and experiences of experts and scientists. Banini and Bearman (1998) implemented FCMs to study and discriminate between the factors affecting the rheological behavior of suspension slurries in the mineral industry. Kok et al. (2000) presented a decision-support system in the field of integrated water management that, with the help of FCMs, is able to incorporate social science scenarios applied to a coastal city. A control system employing FCMs was developed by Kottas et al. (2006) to control the anaerobic digestion processes of a wastewater treatment plant.

4.3   Education

Georgiou and Botsios (2008) presented an adjustable tool for learning style recognition that applies a three layer FCM schema. This extension allows experts and educators to reinforce the learning style recognition of a system and to adjust its accuracy. Hossain and Brooks (2008) proposed a model for educational software adaption based on the perceptions of stakeholders with the use of FCMs. By combining perceptions (i.e., heterogeneous knowledge), FCMs help to facilitate better resource management and thus result in an increased impact on educational software adoption by teachers and students in schools.

4.4   Intelligent Systems

Parenthoen et al. (2001) modeled the emotional behavior of virtual actors (i.e., a shepherd, a dog and a virtual sheep) using FCMs. They showed how autonomous agents can be effectively delocalized in interactive virtual worlds owing to FCMs. Peláez and Bowles (1996) used FCMs in their study to represent and manipulate the type of linguistic knowledge required in failure mode and effect analysis for predicting failure effects and causes in a complex system. Lo Storto (2010) proposed a framework based on FCMs to elicit cognitive schemes and to develop a measure of individual ambiguity tolerance to manage vague situations at the stage of product requirement definition. Another field, Web mining technology, was analyzed by Lee et al. (2002). By suggesting a new concept for web-mining procedures, the authors built on FCMs to transform association rules into FCM-driven causal knowledge bases. Additionally, they attempted to develop a causal knowledge-based inference equivalence property to enrich the interpretation of the draft results of association rule mining. Moreover, Acampora and Loia (2009) researched computational intelligence and developed a novel ambient intelligent architecture integrating FCMs with computational paradigms with the aim of optimizing environmental parameters (e.g., user's comfort and energy savings).

4.5   Medicine

Papageorgiou et al. (2009) employed FCMs to propose a medical decision support for the process of predicting infectious diseases, as well as the type and the severity of infection, with the aim of supporting medical personnel in assessing patient status, making a diagnosis and selecting a therapy. Georgopoulus et al. (2003) proposed a method of differential diagnosis for language impairment, dyslexia and autism by using qualitative and quantitative data, based on accumulated experience and knowledge of human experts, which are represented by FCMs. Some years later, another medical decision support system using FCMs was modeled by Stylios and Georgopoulos (2008) for the differential diagnosis of two language disorders and a speech disorder as well as for obstetrics decision support. Iakovidis and Papageorgiou (2011) applied an intuitionistic FCM to model uncertain, imprecise, and/or incomplete medical knowledge by including the hesitancy of the expert in the determination of the causal relations between the concepts of a specific area. However, FCMs can be used not only for medical-decision making but also for medical diagnosis. Innocent and John (2004) proposed, using FCMs, a new method for computing vague symptoms and temporal information in a clinical diagnostic application to encode fuzzy causal structures. Rodin et al. (2009) implemented FCMs to model and simulate multiple myeloma cells to mimic a simple organism. Furthermore, FCMs were used to represent and compare traditional and scientific knowledge with respect to the causal determinants of diabetes in the study of Giles et al. (2007) to identify more concrete stressors and outcomes that are amenable to management and monitoring.

5 Enhanced Smart Applications Through Learning Algorithms

This section offers an overview of existing smart applications that use FCMs and that are complemented by learning algorithms to be able to understand, reflect and learn from urban processes. For ease of comprehension, the applications are categorized into convenient categories, summarized in Table 2,[6] and shortly reviewed for the final analysis in Sect. 6. A total of 19 research studies (including only the FCM applications using learning algorithms) have been selected, which have been categorized into four fields—business (4), ecology (7), education (3) and medicine (5)—to be presented in this chapter.

5.1 Business

Andreou et al. (2005) used FCMs to model political and strategic issues and situations and to thus support the decision-making process in view of an imminent crisis. In this study, the FCMs are improved by a genetic algorithm that has to find the optimal weight matrix that satisfies a predefined activation level for a certain concept by solving the problem of the invariability of the weights and the inability of the methods to represent a specific political situation following the change of a specific weight(s). Both Kim et al. (2008) and Trappey et al. (2010) proposed an expert system. In the expert system from Kim et al. (2008), FCMs are applied for the forward-backward analysis of the radio frequency identification supply chain, using a genetic algorithm

---

[6]Please note: the left column always states the source, the middle column the applied field and the right the employed learning algorithm.

Table 2 FCM applications with learning algorithms

Business		
Andreou et al. (2005)	Crisis management	Genetic algorithm
Kim et al. (2008)	RFID supply chain	Genetic algorithm
Miao et al. (2007)	Recommendation system	Member function learning algorithm
Trappey et al. (2010)	Logistics	Genetic algorithm
Ecology		
Acampora et al. (2011)	Ambient intelligence	Unsupervised online real-time learning algorithm
Baykasoglu et al. (2011)	Industrial process control	Extended great deluge algorithm
Dickerson and Kosko (1994)	Undersea virtual worlds	Differential Hebbian learning algorithm
Kurgan et al. (2007)	Biology	Genetic algorithm
Lu et al. (2010)	Heating management	Supervised learning algorithm
Stylios and Groumpos (1999)	Manufacturing system	*Implied algorithms*
Xirogiannis et al. (2004)	Urban design	*Implied algorithms*
Education		
Cai et al. (2010)	Game world	Evolutionary algorithm
Luo et al. (2009/2010)	Game-based learning	Hebbian learning algorithm
Medicine		
Beena and Ganguli (2011)	Health monitoring	Activation Hebbian learning algorithm
Froelich and Wakulicz-Deja (2009)	Therapy planning	Balanced differential learning algorithm
Kannappan et al. (2011)	Autism	Unsupervised non-linear Hebbian learning algorithm
Papageorgiou et al. (2006)	Urinary bladder	Unsupervised active Hebbian learning algorithm
Papageorgiou (2011)	Radiation therapy	Nonlinear Hebbian learning algorithm

to minimize the prediction errors made by the forward analysis. The other study, developed by Trappey et al. (2010), used an FCM model to improve a reverse logistic process decision support whose data are collected by radio frequency identification for real-time monitoring. They applied a genetic algorithm to assign weights to the arcs between model nodes to improve the reverse logistic efficiency.

Finally, Miao et al. (2007) used extended FCMs to propose fuzzy cognitive agents (i.e., a new type of personalized recommendation agent) that are able to give personalized suggestions based on the preferences of the current user, common preferences of general users, and expert knowledge. Thus, they can help customers make inferences and decisions through numeric computation instead of symbolic and logic deduction. To normalize the numerical attribute data, the member function learning algorithm was implemented to map real world values to the fuzzy concept state values (i.e., a self-organized map learning mechanism).

5.2 Ecology

Lu et al. (2010) proposed a new control strategy (i.e., intelligent control) based on a FCM model for a district heating network system. Hebbian learning rules and unbalance degrees were used to solve the control problem in a real industrial area. Thus, by applying the supervised learning method, the weights of the FCMs were determined by historical data rather than by experts. Baykasoglu et al. (2011) and Xirogiannis et al. (2004) also addressed control systems. Baykasoglu et al. (2011) proposed an industrial process control based on FCMs to model dynamic systems in which the behavior of the system can be observed quickly. The extended great deluge algorithm was employed as a training algorithm for FCMs to determine the proper weight matrices for the FCM inference through the minimization of a predetermined objective function. Xirogiannis et al. (2004) applied FCMs to model specialist knowledge that contributes to the development of more intelligent control methods and autonomous decision support mechanisms in urban areas and thus supports the process aesthetic analysis of urban areas. In this study, specific algorithms were proposed for interpreting logic-based rules by FCMs, as were specific algorithms and formulas for calculating the values of multi-branch map hierarchies, thus suggesting an algorithm for transforming rules with no disjoints to FCMs. Likewise, Stylios and Groumpos (1999) worked to, in a following step, employ the unsupervised Hebbian learning rule to train FCMs and adjust the causal functions-weights of the interconnections among concepts according to the variation of the concepts. In their study, a hierarchical two-level FCM for modeling the supervision of manufacturing systems was developed and should now be improved with learning algorithms.

Dickerson and Kosko (1994) made use of adaptive FCMs to model an undersea virtual world of dolphins, fish, and sharks. Differential Hebbian learning was used to encode a feeding and chase sequence in an FCM and to avoid spurious causality. In the research from Kurgan et al. (2007), to analyze the secondary structure content of homology proteins, FCMs were applied to quantify the strength (i.e., degree) of the relation between the hydrophobicity scales/indices and the protein content values. A real-coded genetic algorithm was used to allow for automated generation of FCM models from data to perform comparisons between the scales and content values. Acampora et al. (2011) implemented a novel unsupervised online real-time learning algorithm that constructs a fuzzy rule base to improve the pre-built ambient intelligence architecture that is based on a multi-agent system built using FCMs to provide personalized living scenarios by anticipating the behavior of the user, as well as satisfying her/his needs.

5.3   Education

Cai et al. (2010) used evolutionary FCMs to model the dynamic concepts and their causal relationship (including modeling of the characters and the contexts) of a game world. An interactive evolutionary algorithm was used to adjust the weights of the FCM and to determine the causal relationships in the FCM to provide the players with a more engaging and immersive gaming experience. In the study of Luo et al. (2009, 2010), FCMs were used to design a new game-based learning system, integrating a teacher submodel, a learner submodel and a set of learning mechanisms to compute the difference between the outputs of the teacher submodel and the learner submodel, to control the whole learning process based on the differences and to enable the reflection of the learning process of the learner. A Hebbian learning rule was utilized to acquire new knowledge from existing data and to correct false transcendental knowledge. The use of Hebbian learning rules and unbalance degrees makes FCMs more suitable for designing a game-based learning system.

5.4   Medicine

Beena and Ganguli (2011) applied FCMs to detect structural damage in a cantilever beam using measured frequencies. An activation Hebbian learning algorithm was used to improve the classification (i.e., damage detection) ability of the FCMs. This new algorithmic approach represents an enhanced structural health monitoring tool. In the study of Kannappan et al. (2011), FCMs were used for an expert system for modeling and predicting autistic spectrum disorders to improve the medical diagnosis. An unsupervised non-linear Hebbian learning algorithm was employed to train FCMs for the specific problems of autistic disorders to improve their efficiency.

An advanced diagnostic method for urinary bladder tumor grading based on augmented FCMs (i.e., representing specialized knowledge) was developed by Papageorgiou et al. (2006). The unsupervised active Hebbian learning algorithm was used to adjust the weights of the FCM interconnections by adapting all weights for the FCM model to concepts using an acyclic fragment approach and to modify them according to the specific problem characteristics to provide a new advanced diagnostic tool that is highly accurate, transparent and interpretable. Some years later, Papageorgiou (2011) created a framework of augmented FCMs (combining expert knowledge and knowledge from data) to handle the complex problem of making decisions in the radiation therapy decision support system. The nonlinear Hebbian learning algorithm was implemented to train the FCMs to achieve improved decision accuracy and interpretability.

Last, Froelich and Wakulicz-Deja (2009) applied FCMs to discover temporal dependencies between medical concepts for facilitating medical decision support and therapy planning. In their study, they employed the balanced differential learning algorithm. This algorithm was applied to take into account the fact that changes in the effect concept can depend on the existence of a particular activation pattern.

6 Evolution of Cognitive Cities with Learning Processes

As the extensive literature review shows, there already exist many FCM applications that have been envisaged for or implemented in areas relevant to cities. According to Portmann and Finger (2015), a smart city consists of various areas, such as smart democracy, smart education, smart logistics, smart people and smart security, and therefore, the amount of possible FCM applications in these areas is unimaginable. Nevertheless, in comparison to the possibilities that a city offers for the implementation of such applications, research about FCM applied to urban issues is only in its early stages.

Most often, FCMs were employed to connect different sources of knowledge (i.e., domain knowledge and expert knowledge) to facilitate decision making. Thus, a challenge in many fields is in decision making, specifically in the weighing of possible alternatives. Decisions are made based on existing information, and thus, decisions might have been different had more information been available. Especially in precarious situations (e.g., medical emergencies), such an accumulation of knowledge and a simple handling of it are advantageous. Nonetheless, the implementation of FCMs is still quite slow.

The challenge of dealing with natural language (i.e., data in any shape) could be a reason why many researchers avoid this research field, or it could be the fact that calculations that have a certain tolerance toward imprecision and uncertainty are not generally acknowledged (cf. the principal rationales for computing with words [Zadeh 2011]). Whatever the reason, such data often contain the information that would be necessary to further develop one's research or, in this context, a city. A smart city has to overcome the imprecision and uncertainty that are present in natural language. The most important stakeholders of a city (e.g., citizens and out-of-town visitors) perceive, think and express themselves in natural language, which, in turn, suggests that their needs are often ambiguous and difficult to elicit. However, those perceptions and experiences are crucial for the development of a city and thus have to be taken into consideration.

A city can only further develop its processes when it receives feedback from the users (e.g., citizens, companies, and government). In addition, it can only be viewed as a smart city when its citizens utilize the city processes and when they are satisfied with them. That is, processes have to be developed based on the needs of the citizens in order for the processes to be viewed positively. However, it is not easy to elicit perceptions and experiences, and thus, it is essential to apply methods that can handle imprecise data. Without feedback from the city stakeholders (i.e., the users of the urban processes), no evaluation, reflection and improvement of the urban processes and thus of the city are possible. This becomes even more important in the development of cognitive cities because, while a smart city receives feedback from citizens, a cognitive city must learn from them by exchanging information and continuously updating its knowledge. To create an urban intelligence, the unobstructed learning process between city and citizens has to occur, in which perceptions and experiences in natural language play an important role.

When one now includes the learning process according to Siemens (2005), which implies that a collective urban knowledge base can only be created through the exchange of experiences and perceptions of different people, it quickly becomes evident that FCMs by themselves are not able to foster this learning process. As it is exactly this component that allows the transformation of a smart city into a cognitive city, it is important to overcome the drawbacks of FCMs. Although FCMs are a technique with a great potential to use natural language, they still have some shortcomings, namely, a potential uncontrollable convergence to undesired steady states and a critical dependence on experts (Papageorgiou et al. 2004). Thus, FCMs are able to visualize concepts and objects as nodes and to indicate the relationships between these nodes with edges. That is, FCMs allow the display, in a network-like structure, of how pieces of information are linked and what effects they have on each other. However, FCMs, per se, are not able to learn from these data.

When the values of cause-effect weights among concepts are modified, learning algorithms can be helpful for the improvement of the shortcomings of FCMs (Papageorgiou et al. 2004). As mentioned in this chapter, there already exist certain FCM applications that apply learning algorithms. Most often, genetic algorithms and Hebbian learning were mentioned. Genetic algorithms are probabilistic search procedures that are based on natural selection and genetics (Goldberg and Holland 1988; Whitley 1994). Hebbian learning is based on the idea that there should be an increase in the synaptic efficacy between two neurons if they are simultaneously active and a decrease if they are not (Gerstner et al. 1999). The primary feature of learning algorithms, namely, their ability to learn from data by themselves, is very helpful, especially in combination with FCMs and in a city context. Learning algorithms are able to modify the weight matrices of FCMs to include, for example, updates of human expert knowledge or historical data and can therefore improve a city's understanding of real world problems and of the way it copes with them (Kannappan et al. 2011; Papageorgiou 2012).

Even if this seems logical, the practice and the research show that this, too, is still only in the early stages and that it is therefore all the more important to address it. Although Kosko (1986) already started to develop FCMs 30 years ago and implemented certain learning algorithms shortly after that (cf. Dickerson and Kosko [1994]), the development of FCM applications with learning algorithms has progressed slowly, according to our research. However, the analysis also shows that there is a trend toward increasing research, that increasingly more researchers are researching the possibilities of linking FCMs and learning algorithms and that this combination could be a promising way to promote cognitive cities.

7 Conclusions and Outlook

This article shows the current state-of-the-art research in the field of FCMs and FCMs combined with learning algorithms in the context of smart and cognitive cities. For this analysis, 59 of approximately 150 articles were selected and placed into five categories (i.e., business, ecology, education, intelligent systems and medicine). Analyzing these articles in more detail made it possible to understand why FCMs were applied and how learning algorithms in 19 of 59 articles could help to improve FCMs.

One primary advantage of the use of FCMs is that they facilitate the representation of knowledge and the integration of expert knowledge. Learning algorithms, by learning from data, are able to aid a city in better understanding, reflecting and acquiring information from its citizens and from urban processes. Furthermore, it can be stated that learning algorithms have the potential to enhance the ability of FCMs and can be an advantage, especially when applying them during the transformation from smart cities to cognitive cities. However, research in this field is still in its infancy.

The authors are aware that this literature analysis of FCM applications (with or without learning algorithms) in urban fields is not conclusive due to the impossibility of including all research works because they are either not available or not yet published. Nevertheless, this work allows for an analysis of the current state, which can be helpful in fostering the development of cities.

A limitation of the analysis is that it is based only on literature and therefore excludes the mathematical engineering perspective that is especially decisive in cities because of the importance of the feasibility, the reliability and the usability of such tools in regard to a successful implementation.

Additionally, in this article, learning algorithms were only evaluated on a basic level. In a next step, learning algorithms, especially those that were mentioned several times (e.g., genetic algorithms or Hebbian learning) should be addressed in more detail. It should also be determined whether certain algorithms can be especially helpful for urban processes.

Another important aspect, which has not been considered in this article, is the sensitivity of urban data, especially data concerning human beings. It is likely that citizens will not participate in urban learning processes if they are unsure whether the data they provide will be stored safely and if it is not transparent who will have access to the data. Privacy and data security are thus important concepts that cannot be ignored in city development. In future research, there should be a stronger focus on this aspect.

Finally, the treatment of natural language still poses the primary challenge. Especially in consideration of the desirable development of cognitive cities, the soft computing method of computing with words (Zadeh 1996) is of great importance and should be examined more closely in the near future.

References

Acampora G, Loia V (2009) A dynamical cognitive multi-agent system for enhancing ambient intelligence scenarios. In: IEEE International Conference on Fuzzy Systems, 2009. FUZZ-IEEE 2009. IEEE

Acampora G, Loia V, Vitiello A (2011) Distributing emotional services in ambient intelligence through cognitive agents. SOCA 5(1):17–35

Aguilar J (2013) Different dynamic causal relationship approaches for cognitive maps. Appl Soft Comput 13:271–282

Altay A, Kayakutlu G (2011) Fuzzy cognitive mapping in factor elimination: a case study for innovative power and risks. Procedia Comput Sci 3:1111–1119

Andreou AS, Mateou NH, Zombanakis GA (2005) Soft computing for crisis management and political decision making: the use of genetically evolved fuzzy cognitive maps. Soft Comput 9(3):194–210

Banini GA, Bearman RA (1998) Application of fuzzy cognitive maps to factors affecting slurry rheology. Int J Miner Process 52(4):233–244

Baykasoglu A, Durmusoglu ZDU, Kaplanoglu V (2011) Training fuzzy cognitive maps via extended great deluge algorithm with applications. Comput Ind 62(2):187–195

Beena P, Ganguli R (2011) Structural damage detection using fuzzy cognitive maps and Hebbian learning. Appl Soft Comput 11(1):1014–1020

Bertolini M (2007) Assessment of human reliability factors: a fuzzy cognitive maps approach. Int J Ind Ergon 37(5):405–413

Cai Y, Miao C, Tan AH, Shen Z, Li B (2010) Creating an immersive game world with evolutionary fuzzy cognitive maps. IEEE Comput Graph Appl 30(2):58–70

Campbell T (2009) Learning cities: knowledge, capacity and competitiveness. Habitat Int 33(2):195–201

Chen H, Chiang RH, Storey VC (2012) Business intelligence and analytics: from big data to big impact. MIS Q 36(4):1165–1188

Chytas P, Glykas M, Valiris G (2011) A proactive balanced scorecard. Int J Inf Manage 31(5):460–468

Çoban O, Seçme G (2005) Prediction of socio-economical consequences of privatization at the firm level with fuzzy cognitive mapping. Inf Sci 169(1):131–154

Dickerson JA, Kosko B (1994) Virtual worlds as fuzzy cognitive maps. Presence Teleoperators Virtual Environ 3(2):173–189

D'Onofrio S, Portmann E (2015) Von Fuzzy-Sets zu Computing-with-Words. Informatik-Spektrum 38:1–7

D'Onofrio S, Portmann E, Kaltenrieder P, Myrach T (2016) Enhanced knowledge management by synchronizing mind maps and fuzzy cognitive maps. In: International conference on fuzzy management methods. Fribourg, Switzerland

D'Onofrio S, Wehrle M, Portmann E (2017) Striving for semantic convergence. In: IEEE International Conference on Fuzzy Systems (FUZZ-IEEE 2017), Naples, Italy (Submitted)

Downs RM, Stea D (1973) Cognitive maps and spatial behavior. Image and environment. Aldine Publishing Company, Chicago

Froelich W, Wakulicz-Deja A (2009) Mining temporal medical data using adaptive fuzzy cognitive maps. In: 2nd conference on human system interactions, 2009. HIS 2009. IEEE

Georgopoulos, Voula C, Georgia A Malandraki, Chrysostomos D Stylios (2003) A fuzzy cognitive map approach to differential diagnosis of specific language impairment. Artif intell Med 29(3):261–278

Georgiou DA, Botsios SD (2008) Learning style recognition: a three layers fuzzy cognitive map schema. In: IEEE international conference on fuzzy systems, 2008. FUZZ-IEEE 2008. IEEE (IEEE World Congress on Computational Intelligence)

Gerstner W, Kempter R, van Hemmen JL, Wagner H (1999) Pulsed neural networks, chapter hebbian learning of pulse timing in the barn owl auditory system. Bradford Books, MIT Press, Cambridge

Giles BG, Findlay CS, Haas G, LaFrance B, Laughing W, Pembleton S (2007) Integrating conventional science and aboriginal perspectives on diabetes using fuzzy cognitive maps. Soc Sci Med 64(3):562–576

Goldberg DE, Holland JH (1988) Genetic algorithms and machine learning. Mach Learn 3(2):95–99

Goodchild MF (2007) Citizens as Sensors: Web 2.0 and the volunteering of geographic information. GeoFocus (Editorial) 7:8–10

Hanafizadeh P, Aliehyaei R (2011) The application of fuzzy cognitive map in soft system methodology. Syst Pract Action Res 24(4):325–354

Hobbs JR (1985) Granularity. In: Proceedings of international joint conference on artificial intelligence (IJCAI), Los Angeles, CA, pp 432–435

Hossain S, Brooks L (2008) Fuzzy cognitive map modelling educational software adoption. Comput Educ 51(4):1569–1588

Hurwitz JS, Kaufman M, Bowles A (2015) Cognitive computing and big data analytics. Wiley, Hoboken, New Jersey

Iakovidis DK, Papageorgiou E (2011) Intuitionistic fuzzy cognitive maps for medical decision making. IEEE Trans Inf Technol Biomed 15(1):100–107

Innocent PR, John RI (2004) Computer aided fuzzy medical diagnosis. Inf Sci 162(2):81–104

Irani Z, Sharif A, Love PE, Kahraman C (2002) Applying concepts of fuzzy cognitive mapping to model: the IT/IS investment evaluation process. Int J Prod Econ 75(1):199–211

Jetter A, Schweinfort W (2011) Building scenarios with Fuzzy Cognitive Maps: an exploratory study of solar energy. Futures 43(1):52–66

Kaltenrieder P, D'Onofrio S, Portmann E (2015) Enhancing multidirectional communication for cognitive cities. In: 2nd international conference on eDemocracy & eGovernment (ICEDEG). IEEE, pp 38–43

Kaltenrieder P, Altun T, D'Onofrio S, Portmann E, Myrach T (2016) Personal digital assistant 2.0 –a software prototype for cognitive cities. In: IEEE international conference on fuzzy systems (FUZZ-IEEE 2016), Vancouver, Canada

Kang I, Lee S, Choi J (2004) Using fuzzy cognitive map for the relationship management in airline service. Expert Syst Appl 26(4):545–555

Kannappan A, Tamilarasi A, Papageorgiou EI (2011) Analyzing the performance of fuzzy cognitive maps with non-linear hebbian learning algorithm in predicting autistic disorder. Expert Syst Appl 38(3):1282–1292

Kardaras D, Karakostas B (1999) The use of fuzzy cognitive maps to simulate the information systems strategic planning process. Inf Softw Technol 41(4):197–210

Kelly III JE (2015) Computing, cognition and the future of knowing. How humans and machines are forging a new age of understanding. IBM Research

Kim MC, Kim CO, Hong SR, Kwon IH (2008) Forward-backward analysis of RFID-enabled supply chain using fuzzy cognitive map and genetic algorithm. Expert Syst Appl 35(3):1166–1176

Kok K (2009) The potential of fuzzy cognitive maps for semi-quantitative scenario development, with an example from Brazil. Glob Environ Change 19(1):122–133

Kok JL, Titus M, Wind HG (2000) Application of fuzzy sets and cognitive maps to incorporate social science scenarios in integrated assessment models. A case study of urbanization in Ujung Pandang, Indonesia. Integr Assess 1(3):177–188

Kosko B (1986) Fuzzy cognitive maps. Int J Man Mach Stud 24(1):65–75

Kottas T, Boutalis Y, Diamantis V, Kosmidou O, Aivasidis A (2006) A fuzzy cognitive network based control scheme for an anaerobic digestion process. In: 14th Mediterranean conference on control and automation, 2006. MED 2006. IEEE

Kurgan LA, Stach W, Ruan J (2007) Novel scales based on hydrophobicity indices for secondary protein structure. J Theor Biol 248(2):354–366

Laird L (1919) The law of parsimony. Monist 29(3):321–344

Lazzerini B, Mkrtchyan L (2010) Risk analysis using extended fuzzy cognitive maps. In: 2010 international conference on intelligent computing and cognitive informatics (ICICCI). IEEE

Lee KC, Kim JS, Chung NH, Kwon SJ (2002) Fuzzy cognitive map approach to web-mining inference amplification. Expert Syst Appl 22:197–211

Lee S, Kim BG, Lee K (2004) Fuzzy cognitive map-based approach to evaluate EDI performance: a test of causal model. Expert Syst Appl 27(2):287–299

Lo Storto C (2010) Assessing ambiguity tolerance in staffing software development teams by analyzing cognitive maps of engineers and technical managers. In: 2nd international conference on engineering systems management and its applications (ICESMA), 2010. IEEE

Lu W, Yang J, Li Y (2010) Control method based on fuzzy cognitive map and its application on district heating network. In: International conference on intelligent control and information processing (ICICIP), 2010. IEEE

Luo X, Wei X, Zhang J (2009) Game-based learning model using fuzzy cognitive map. In: Proceedings of the first ACM international workshop on multimedia technologies for distance learning. New York, ACM

Luo X, Wei X, Zhang J (2010) Guided game-based learning using fuzzy cognitive maps. IEEE Trans Learn Technol 3(4):344–357

Malone TW, Bernstein MS (2015) Handbook of collective intelligence. MIT Press, Cambridge

Miao C, Yang Q, Fang H, Goh A (2007) A cognitive approach for agent-based personalized recommendation. Knowl-Based Syst 20(4):397–405

Mostashari A, Arnold F, Mansouri M, Finger M (2011) Cognitive cities and intelligent urban governance. Netw Ind Q 13(3):4–7

Papageorgiou EI (2011) A new methodology for decisions in medical informatics using fuzzy cognitive maps based on fuzzy rule-extraction techniques. Appl Soft Comput 11(1):500–513

Papageorgiou EI (2012) Learning algorithms for fuzzy cognitive maps—a review study. IEEE Trans Syst Man Cybern Part C Appl Rev 42(2)

Papageorgiou EI, Salmeron JL (2013) A review of fuzzy cognitive maps research during the last decade. IEEE Trans Fuzzy Syst 21(1):66–79

Papageorgiou EI, Stylios CD, Groumpos PP (2004) Active Hebbian learning algorithm to train fuzzy cognitive maps. Int J Approx Reason 37(3):219–249

Papageorgiou EI, Stylios CD, Groumpos PP (2006) Unsupervised learning techniques for fine-tuning fuzzy cognitive map causal links. Int J Hum Comput Stud 64(8):727–743

Papageorgiou EI, Stylios CD, Groumpos PP (2009) A fuzzy cognitive map based tool for prediction of infectious diseases. In: IEEE international conference on fuzzy systems, 2009. FUZZ-IEEE 2009. IEEE

Parenthoen M, Reignier P, Tisseau J (2001) Put fuzzy cognitive maps to work in virtual worlds. In: The 10th IEEE international conference on fuzzy systems, 2001, vol 1. IEEE

Peláez CE, Bowles JB (1996) Using fuzzy cognitive maps as a system model for failure modes and effects analysis. Inf Sci 88(1–4):177–199

Portmann E, Finger M (2015) Smart Cities–Ein Überblick! HMD Praxis der Wirtschaftsinformatik 1–12

Rodin V, Querrec G, Ballet P, Bataille FR, Desmeulles G, Abgrall JF, Tisseau J (2009) Multi-agents system to model cell signalling by using fuzzy cognitive maps. Application to computer simulation of multiple myeloma. In: 9th IEEE international conference on bioinformatics and bioengineering, 2009. BIBE 2009. IEEE

Rodriguez-Repiso L, Setchi R, Salmeron JL (2007) Modelling IT projects success with fuzzy cognitive maps. Expert Syst Appl 32:543–559

Saaty TL (1996) Decision making with dependence and feedback: the analytic network process. RWS Publications, Pittsburgh

Salmeron JL (2009) Supporting decision makers with fuzzy cognitive maps. Res Technol Manag 52(3):53–59

Salmeron JL, Lopez C (2012) Forecasting risk impact on ERP maintenance with augmented fuzzy cognitive maps. IEEE Trans Softw Eng 38(2):439–452

Sharif AM, Irani Z (2006) Exploring fuzzy cognitive mapping for IS evaluation. Eur J Oper Res 173(3):1175–1187

Siemens G (2005) Connectivism: a learning theory for the digital age. Int J Instr Technol Distance Learn 2(1):3–10

Skov F, Svenning JC (2003) Predicting plant species richness in a managed forest. For Ecol Manage 180:583–593

Strauss A, Corbin J (1994) Grounded theory methodology. Handb Qual Res 17:73–85

Stylios CD, Groumpos PP (1999) A soft computing approach for modelling the supervisor of manufacturing systems. J Intell Rob Syst 26(3):389–403

Stylios CD, Georgopoulos VC (2008) Fuzzy cognitive maps structure for medical decision support systems. Forging new frontiers. Fuzzy pioneers II. Springer, Berlin

Tan CO, Özesmi U (2006) A generic shallow lake ecosystem model based on collective expert knowledge. Hydrobiologica 563(1):125–142

Tolman EC (1948) Cognitive maps in rats and men. Psychol Rev 55(4):15–64

Trappey AJC, Trappey CV, Wu CR (2010) Genetic algorithm dynamic performance evaluation for RFID reverse logistic management. Expert Syst Appl 37(11):7329–7335

van Vliet M, Kok K, Veldkamp T (2010) Linking stakeholders and modellers in scenario studies: the use of fuzzy cognitive maps as a communication and learning tool. Futures 42(1):1–14

Webster J, Watson RT (2002) Analyzing the past to prepare for the future: writing a literature review. MIS quarterly

Whitley D (1994) A genetic algorithm tutorial. Stat Comput 4(2):65–85

Xirogiannis G, Glykas M (2007) Intelligent modeling of e-business maturity. Expert Syst Appl 32(2):687–702

Xirogiannis G, Stefanou J, Glykas M (2004) A fuzzy cognitive map approach to support urban design. Expert Syst Appl 26(2):257–268

Yaman D, Polat S (2009) A fuzzy cognitive map approach for effect-based operations: an illustrative case. Inf Sci 179(4):382–403

Yu R, Tzeng GH (2006) A soft computing method for multi-criteria decision making with dependence and feedback. Appl Math Comput 180(1):63–75

Zadeh LA (1965) Fuzzy sets. Inf Control 8(3):338–353

Zadeh LA (1988) Fuzzy logic. Computer 21(4):83–93

Zadeh LA (1996) Fuzzy logic = computing with words. IEEE Trans Fuzzy Syst 4(2):103–111

Zadeh LA (1998) Some reflections on soft computing, granular computing and their roles in the conception, design and utilization of information/intelligent systems. Soft Comput 2:23–25

Zadeh LA (2006) From search engines to question answering systems—the problems of world knowledge, relevance, deduction and precisiation. In: E. Sanchez (ed) Fuzzy logic and the semantic web, pp 163–210

Zadeh LA (2008) Is there a need for fuzzy logic? Inf Sci 178:2751–2779

Zadeh LA (2011) Computing with words—principal concepts and ideas. Studies in fuzziness and soft computing. Springer, Heidelberg

Sara D'Onofrio is Ph.D.-student in Computer Science at the Human-IST Institute of the University of Fribourg, Switzerland, and IT trainee at Swiss Post. She has a bilingual bachelor's degree in Business Administration from the University of Fribourg and a master's degree in Business Administration with a specialization in Business Information Systems from the University of Bern, Switzerland. She has attended courses in business modeling (University of Vienna, Austria), fuzzy logic (Lake Como School of Advanced Studies, Italy; Centro Singular de Investigacion en Tecnoloxias da Informacion, Spain), fuzzy cognitive maps (Volos, Greece) and design science and design thinking (Ovronnaz, Switzerland). She took part in several conferences in Europe, South America and Canada. She has written more than 20 articles in journals, editorial books and conference proceedings and is student member of EUSFLAT and IEEE. Her current research interests lie in soft and cognitive computing, natural language processing, and smart and cognitive cities.

Elpiniki I. Papageorgiou is Associate Professor at the Department of Electrical Engineering at University of Applied Sciences (TEI) of Thessaly, Larisa, Greece. She holds a Ph.D. in Computer Science from the University of Patras (2004). She is a specialist in developing and applying soft computing methods and algorithms to decision support problems for prediction, strategic decisions, scenario analysis and data mining. Her main research field is the development of novel algorithms and methods for intelligent decision support systems and Fuzzy Cognitive Maps. She has been working as principal investigator and senior researcher in several research projects related with the development of novel computational intelligence methodologies for decision support systems, intelligent algorithms and fuzzy cognitive tools for prediction and decision making, data mining, big data analysis (participation as Researcher/Technical Manager in 13 EU and several national projects). Today, she is PI in four national research projects at the Center of Technology Research Stereas Elladas concerning the development of intelligent algorithms for data analysis and DSS in energy and water. She has more than 195 publications in journals, conference papers and book chapters and is the editor of "Fuzzy Cognitive Maps for Applied Sciences and Engineering—From Fundamentals to Extensions and Learning Algorithms" (2014). Elpiniki Papageorgiou is a reviewer in many international journals, IEEE Senior Member and member in IEEE CIS. Her research interests include intelligent systems, fuzzy cognitive maps, soft computing methods, decision support systems, cognitive systems, data mining and machine learning.

Edy Portmann is a researcher and scholar, specialist and consultant for semantic search, social media, and soft computing. Currently, he works as a Swiss Post-Funded Professor of Computer Science at the Human-IST Institute of the University of Fribourg, Switzerland. Edy Portmann studied for a BSc in Information Systems at the Lucerne University of Applied Sciences and Arts, for an MSc in Business and Economics at the University of Basel, and for a Ph.D. in Computer Sciences at the University of Fribourg. He was a Visiting Research Scholar at National University of Singapore (NUS), Postdoctoral Researcher at University of California at Berkeley, USA, and Assistant Professor at the University of Bern. Next to his studies, Edy Portmann worked several years in a number of organizations in study-related disciplines. Among others, he worked as Supervisor at Link Market Research Institute, as Contract Manager for Swisscom Mobile, as Business Analyst for PwC, as IT Auditor at Ernst & Young and, in addition to his doctoral studies, as Researcher at the Lucerne University of Applied Science and Arts. Edy Portmann is repeated nominee for Marquis Who's Who, selected member of the Heidelberg Laureate Forum, co-founder of Mediamatics, and co-editor of the Springer Series 'Fuzzy Management Methods', as well as author of several popular books in his field. He lives happily married in Bern and has three lively kids.

The Role of Interpretable Fuzzy Systems in Designing Cognitive Cities

José M. Alonso, Ciro Castiello and Corrado Mencar

Abstract In recent years, there has been a huge effort connecting all kind of devices to Internet. From small devices (e.g., e-health monitoring sensors or mobile phones) that we carry daily in what is called the body-area-network, to big devices (such as cars), passing by all devices (e.g., TVs or refrigerators) at home. In modern cities, everything (at work, at home, and even in the streets) is connected to Internet. Accordingly, the amount of data in Internet grows dramatically every day. With this regard, humans face two main challenges: (1) to extract valuable knowledge from the given Big Data and (2) to become part of the equation, i.e., to become active actors in the Internet of Things. To do so, researchers and developers have created a novel generation of intelligent systems which are producing more and more intelligent devices, yielding what is called smart cities. Fuzzy systems are used in many applications in the context of Smart Cities. Now, it is time to address the effective interaction between intelligent systems and citizens with the aim of passing from smart to Cognitive Cities. Moreover, the use of interpretable fuzzy systems can facilitate such interaction and pave the way towards Cognitive Cities.

Keywords Interpretability · Fuzzy Logic
Computational Theory of Perceptions
Linguistic descriptions of complex phenomena · Collaborative intelligence
Human-machine communication
Cognitive City

J. M. Alonso (✉)
Centro Singular de Investigación en Tecnoloxías da Información (CiTIUS),
Universidade de Santiago de Compostela, Rúa de Jenaro de la Fuente Domínguez,
15782 Santiago de Compostela, Spain
e-mail: josemaria.alonso.moral@usc.es

C. Castiello · C. Mencar
Department of Informatics, University of Bari "Aldo Moro", v. E. Orabona n. 4,
70125 Bari, Italy
e-mail: ciro.castiello@uniba.it

C. Mencar
e-mail: corrado.mencar@uniba.it

E. Portmann et al. (eds.), *Designing Cognitive Cities*, Studies in Systems,
Decision and Control 176, https://doi.org/10.1007/978-3-030-00317-3_6

1   Introduction

According to the 2014 revision of United Nations's World Urbanization Prospects (United Nations and Social Affairs 2014), 54% of the world's population lives in urban areas. This proportion is expected to increase to 66%. Therefore cities are becoming more complex, especially in terms of information density, thus calling for complex ICT solutions in order to make cities more competent and functional. Apparently, as the complexity increases, ICT enables the emergence of

> behaviors and characteristics that are both unpredictable and potentially quite powerful and ones that occur without passing through human institutions or filters, generating urban 'culture'

(i.e., patterns of human activity and the symbolic structures that give such activity significance) (Tusnovics 2007). The "urban cultural" design of cities requires new approaches that could be borrowed by the Cognitive Sciences (Howard 1983), and which suggest the idea of "Cognitive City".

The concept of "Cognitive City" was first introduced by Novak (1997) and later refined by other authors. There are affine definitions concerning the rethinking of a city as an intelligent, information-centric environment, such as "Smart City", "Intelligent City", etc. Namely, a Cognitive City emphasizes the role of learning, memory creation and experience retrieval as central processes for coping with current challenges of efficiency, sustainability and resilience (Finger and Portmann 2016; Mostashari et al. 2011).

As any other cognitive system, a Cognitive City operates in accordance with mechanisms of perception, processing (association, planning, choice, etc.), actualization and feedback (Mostashari et al. 2011). Peculiar to Cognitive Cities, however, is the twofold role of citizens who are "sensors" as well as "recipients". In other terms, people feed a cognitive system with information and knowledge; at the same time, people learn knowledge from a Cognitive City too (Finger and Portmann 2016).

In a nutshell, an intelligent and distributed collaboration takes place between the Cognitive City and its citizens. This appeals for a novel form of intelligence, which may be called "collaborative intelligence", where people and machines collaborate to solve complex problems (Epstein 2015). As a matter of fact, research on collaborative systems is a hot trend for the new Artificial Intelligence (AI) (Russell and Norvig 2003) in the coming years (Stone et al. 2016). Notice that effective human-machine communication requires mutual understanding. Moreover, the USA Defense Advanced Research Projects Agency (DARPA) has recently remarked that the effectiveness of current AI systems is jeopardized because of their lack of explainability when interacting with humans (Gunning 2016). Thus, DARPA is looking for new explainable AI systems (XAI systems in short), i.e., systems endowed with the capability to be understood and trusted on.

Accordingly, to collaborate effectively with people, a Cognitive City must be populated with XAI systems able to model the human view of the world (Epstein 2015). The required information/knowledge exchange is more effective if a suitable medium for knowledge representation is adopted. In fact, whenever actors operate this communication with different languages, incomprehension and social exclusion may arise (Perticone and Tabacchi 2016).

What is assumed here is that people acquire information through perceptions (i.e., the organization, identification, and interpretation of sensory information—or sensations—to form mental representations Schacter et al. 2015) and build their common-sense knowledge through them. Consequently, mental conceptualizations are basically perceptual in nature. Therefore any cognitive system aiming at exchanging information and knowledge with people should confront with the ability of storing and processing perception-based information. Thus, it is required a step ahead in AI with the aim of properly handling both computational and human perceptions (Zadeh 2001). Notice that human perceptions are defined by intrinsic and extrinsic attributes regarding human senses (sight, smell, taste, touch and hearing). In addition, human pleasantness depends on what people experience (Perceptions), but also on what people expect (Cognitions) which is influenced by common and personal background (context, mood, etc.).

Fuzzy Logic (in the wide sense) is a general paradigm that enables the representation and processing of perceptual information as opposed to the traditional way of computing measurements (Zadeh 1999). It is a logic for dealing with imprecision and approximate reasoning (Zadeh 2008), which are two key ingredients for manipulating perceptive information. This is accomplished by shifting from the classical idea of information as represented by a precise value (i.e., measurement-based) to a *granular* representation of information, where many values are put together according to their similarity, proximity, etc., and considered as an integrated whole (Pedrycz et al. 2008). Fuzzy Logic applications originated in the realm of intelligent control, then they spread in expert systems, optimization, decision making, etc. (Yager and Zadeh 2012). In addition, Zadeh introduced the Computational Theory of Perceptions (CTP) (Zadeh 1999, 2002) with the aim of facilitating the generation of AI systems ready to compute with imprecise descriptions of the world in a similar way how humans naturally do. Other authors have extended and used CTP in different applications, e.g., data mining (Yager 1995), database queries (Kacprzyk and Zadrozny 2010), or description of temporal series (Al-Hmouz et al. 2015).

Given these premises, it is not surprising that Fuzzy Logic finds a natural application to Cognitive Cities (Wilke and Portmann 2016). The CTP is rooted in the Fuzzy Set Theory (FST) which extends the classical concept of 'set' in order to accommodate partial membership (Zadeh 1965). This extension gives rise to an extreme flexibility of models and methods based on FST. Nevertheless, this flexibility implies a serious drawback. In fact, models are usually built according to automatic processes of inductive learning, which are aimed at optimizing some performance criteria. In most cases emphasis is put on the accuracy of the model with respect to some standard. Automatic learning is a necessity whenever human intervention in model design is limited or impossible due to the high complexity of the problem to be tackled, which is usually manifested in large amounts of high-dimensional and heterogeneous data. This problem is further exacerbated by the emergence of big data (Domingos 2012). Nevertheless, in consequence of an unconstrained learning toward the most accurate solutions, the flexibility of FST could easily lead to resulting models that are far from any possible attempt of interpreting the acquired knowledge in terms of common-sense concepts as those possessed by people (especially those with non-technical skills).

A problem of *interpretability* thus arises when the design of a Fuzzy Logic model is made by a non-meticulous designer or committed to some automatic learning process (Alonso et al. 2015). In order to tackle this problem, the concept of inter-pretability has been first analyzed and decomposed in its structural and semantic facets. Then, a number of interpretability constraints and criteria can be formulated in order to guide the design process toward interpretable fuzzy models that also exhibit good accuracy. Notice that interpretability constraints limit the flexibility of FST, therefore a proper balance between interpretability and other performance metrics must be expected. This is accomplished by proper metrics for assessing inter-pretability, which can be plunged into learning methods to constrain the way models are designed from data.

Interpretable fuzzy systems provide a granular view of their embodied knowledge, which is highly co-intensive with common-sense knowledge possessed by people. In this sense, these systems enable *intuitive legibility* of the data from which they have been designed, thus supporting Human-Data Interaction (Mortier et al. 2014), a necessary condition for enabling the collective intelligence amplification and, therefore, the emergence and growth of Cognitive Cities (Wilke and Portmann 2016). Moreover, interpretable fuzzy systems can be naturally embedded into the so-called Data-to-Text (D2T) systems (Ramos-Soto et al. 2016b; Reiter and Dale 2000; Trivino and Sugeno 2013) which are aimed at describing data coming out of everyday complex phenomena in natural language, as an effective user-friendly alternative to the usual tables and graphs.

The rest of the manuscript is organized as follows. Section 2 introduces some preliminary concepts that are required to understand the rest of the chapter. Section 3 presents a general framework to generate linguistic descriptions of complex phe-nomena in Cognitive Cities. This framework is grounded on the CTP and supported by interpretable fuzzy systems. Section 4 shows an illustrative use case. Finally, we remark the main conclusions and sketch future work in Sect. 5.

2 Preliminaries

2.1 Fuzzy Rule-Based Systems

Fuzzy Rule-Based Systems (FRBSs) represent an extension of classical rule-based systems, relying on the adoption of fuzzy sets as the basic constituting elements. The concept of fuzzy set is characterized by a *membership function*, i.e. a mapping from a Universe of Discourse (UoD) U to a completely distributive lattice (usually, the interval [0, 1]). Formally:

$$\mu_A : U \mapsto [0, 1]$$

is the membership function of the fuzzy set A. Thus,

$$A = \{(x, u) | x \in U, u = \mu_A(x)\}$$

and $\mu_A(x)$ is called the *membership degree* of x to A.

The inherent nature of fuzzy sets provides a way to effectively represent concepts which are commonly manipulated by humans who make use of linguistic expressions. In fact, the information embedded in a fuzzy set is related to a meaning determined by the degree of the membership function.

The understanding of fuzzy sets in terms of linguistic concepts paves the way for a modeling process oriented to linguistically organize the knowledge. This is accomplished by reviewing the fuzzy implication $A \rightarrow B$ as a rule of the form: IF A THEN B, where A and B are fuzzy sets describing the linguistic terms involved in the antecedent and consequent parts of the rule, respectively. The fuzzy rule, therefore, stands as an atomic proposition where a linguistic variable (whose instances are the elements of the UoD) is tight to the linguistic terms specified by the fuzzy sets. The formal description of a rule is as follows:

$$\text{IF } X_1 \text{ is } A_1 \text{ and } \ldots \text{ and } X_n \text{ is } A_n \text{ THEN } Y \text{ is } B. \tag{1}$$

A FRBS is generally composed by a *knowledge base* (containing the knowledge about the problem to be tackled) and an *inference engine* (providing the operations to carry on the IF-THEN inference mechanism). The knowledge base includes a number of rules expressed as in (1) whose definition relies on the fuzzy partitions imposed on the involved UoDs, i.e., the collections of fuzzy sets designed to provide an ensemble of linguistic terms referable to a linguistic variable. The knowledge base, therefore, includes the membership functions of the involved fuzzy sets. On the other hand, the inference engine specifies the steps of the fuzzy reasoning required to derive the output value from the input relationships described by the antecedent parts. In particular, the most common inference mechanism is firstly oriented to derive the output of each rule by extending the fuzzy implication to the multidimensional input of the single premises. Then individual output values are properly aggregated by means of specific operators.

There exist a number of different implementations of FRBSs which differ in the modeling choices involving both the compilation of the knowledge base (with determinations concerning the shape of the fuzzy sets, the definition of fuzzy partitions, the organization of the fuzzy rule base, and so on) and the set-up of the inference engine (where different fuzzy operators may be adopted as well as different interpretations concerning the fuzzy implication, the rules' aggregation, the output specification, etc.) (Zadeh 2011).

FRBSs provide additional evidence in several applications where users need support for diagnoses. The medical context represents an ideal field of reference in this sense (Alonso et al. 2012). It is noteworthy the possibility to employ human experts' knowledge in the form of linguistic variables and rules. In addition, when a suitable amount of data is available, a FRBS can be obtained directly from data. A plethora of proposals are reported in literature in this direction (Pedrycz 2012), which are mainly based on the idea of deriving fuzzy rules as a result of the application of some algorithms related to data clustering or feature space partitioning. In this context, the challenge consists in conjugating the accuracy of the resulting model with the interpretability of the overall inference system. In fact, the possibility for the human user to read and understand the knowledge embedded in a FRBS is a key-issue in fuzzy modeling, as we are going to discuss in the following section.

2.2   Interpretable Fuzzy System Modeling

Fuzzy systems gained attention in the scientific community mainly by virtue of their expressive capabilities. Without renouncing to the strictness of a scientific formalization, fuzzy systems are able to model complex phenomena in a comprehensible form, by resorting to the employment of the natural language spoken by the final users. That is the reason why the catchword "Computing with words" has been frequently associated to this kind of approaches (Kacprzyk and Zadrozny 2010).

When conceiving the application of tools to support human activities in everyday tasks from real-world contexts, the benefits deriving from the employment of fuzzy models are manifold. A knowledge base realized on the basis of fuzzy rules, for example, can be easily compiled, fine-tuned or integrated by humans, especially by experts who are aware of the knowledge domain and understand the expressed relations among data. The use of natural language represents a common ground for knowledge representation and users are therefore invited to a closer interaction with the proposed model. This, in turn, fosters a deeper exploration of the model which leads to the simplification and/or completion of the knowledge representation. Such an activity may prove to be useful even to detect anomalies and inconsistencies which could be plausible from a mathematical point of view, while being plainly exposed when put to the test of common sense. Most importantly, fuzzy models act even as sources of persuasion. In fact, the outcomes produced by a fuzzy rule base can be accepted by users who are convinced by the explanation of the inference process and remain confident about the support provided to their jobs.

However, a common misunderstanding must be avoided: the simple adoption of FST cannot imply *ipso facto* the guarantee to produce interpretable representations of knowledge. In many cases, automatic generation of fuzzy rules results in complex and/or indecipherable models, especially when greater emphasis is laid on accuracy. Interpretability issues, instead, must be deeply analyzed and discussed since comprehension is attained only by means of careful design (Alonso et al. 2015).

From a semantic point of view, it can be argued that the interpretability process is carried on by humans engaged in reading and understanding the knowledge expressed by a fuzzy system. To successfully accomplish such a goal, some underlying requirements must be fulfilled. Namely, we observe that the linguistic variables involved in a fuzzy rule are related to linguistic terms to constitute the input/output components of the inference engine. The linguistic terms find expression in a mathematical formalization—namely, the fuzzy sets—which stand at the basis of the model engine. At the same time, the linguistic terms are also associated to the meaningful representations—namely, the concepts—which stand at the basis of the users' cognition. In this way, concepts and fuzzy sets find a point of convergence represented by the use of common linguistic terms. Such a basic property can be referred to as a semantic *cointension* between concepts and fuzzy sets (Mencar et al. 2011; Zadeh 2008), ensuring the possibility to perform the interpretability process. More precisely, we state that cointension may be related to one facet of interpretability, which is *comprehensibility* of the actual informative contents of a fuzzy knowledge base.

Among the other facets to be considered, the role of the brain's cognitive capabilities (and their cognitive limitations) cannot be neglected. The interpretability process, in fact, is grounded also on the users' capability to contain the bulk of information a fuzzy system can convey. In this sense, simplicity stands as a key factor to allow plain reading and understanding of information: this is the reason why we state that the structural complexity of a model may be related to another facet of interpretability, which is *readability* of the amount of information.

Bearing in mind the above considerations, the interpretability process needs also to be assessed by means of quantitative criteria which can be directly applied on constitutive elements of a fuzzy system.

The basic elements of a fuzzy system are represented by fuzzy sets and their mathematical counterpart, i.e., the membership functions. In principle, any kind of membership functions may be admitted in a modeling process, yet they are ultimately addressed at denoting linguistic concepts. In this sense, the corresponding fuzzy sets should abide by several constraints, including normality (full membership degree ensured by at least one element inside the UoD), continuity (to adhere to the natural way the human perception takes place), and convexity (so that a property satisfied by two elements of the UoD is satisfied also by any other element included among them).

Fuzzy partitions define relationships among a number of fuzzy sets which collectively provide the interpretation of a linguistic term. Several criteria can be proposed to construct interpretable fuzzy partitions: they should include a limited number of fuzzy sets; they should preserve intuitive relationship characterizing the involved concepts (such as the order relation); they should take into account that the UoD may include special elements deserving prototypical roles for some fuzzy sets. Moreover, the overlapping degree of fuzzy sets in a fuzzy partition must be carefully arranged. On the one hand, overlapping should be restrained (in order to ensure the representation of distinguishable concepts). On the other hand, coverage of the UoD must be ensured (so that every element is assigned with a suitable membership function for at least one fuzzy set).

Rules are the means to express the input/output behavior of a fuzzy system. They are composed of atomic propositions linking linguistic variables to linguistic terms, which altogether embody the premise and the consequent of each rule. At this level, the information conveyed by a fuzzy model is compiled in natural language and the users can examine the elements of the inference engine by simply reading them. Such a task may be eased if the structure of each rule is kept simple. In this sense, a reduced number of atomic propositions should be included in fuzzy rules. This kind of criteria aims at reducing the description length of the rules, thus enhancing their readability. Moreover, the output of rules should encapsulate the inherent vagueness that the fuzzy system is going to manage. Therefore, as an additional criterion, granular output should be preferred, described in form of fuzzy sets.

The ensemble of rules composing the fuzzy rule base is at the highest level of knowledge representation. In order to compile readable information, the fuzzy rule base should be composed by a limited number of rules. Such a structural constraint must be carefully adopted since, as a side effect, it generally implies a reduced performance of the inference engine in terms of accuracy. Some other criteria pertaining to semantic constraints may be considered when building up a fuzzy rule base, including completeness (that is proper activation of rules for each proposed input in terms of firing strength) and locality (that is restrained overlapping of rule activity in terms of simultaneous high values of firing strength). Additionally, it should be observed that the IF-THEN structure of fuzzy rule bases refers to the arrangement of logical propositions. Therefore, the reader expects to recover inside the linguistic description of rules the same basic laws of reasoning which drive human logical inference. In this sense, the fuzzy rule base should be able to express a logical view coherently with propositional logic, so that the production of nonsensical results can be avoided. With that aim, it becomes a critical task the right selection/definition of meaningful fuzzy operators involved in the fuzzy inference process.

Notice that interpretable fuzzy models built in accordance with the guidelines described above are ready to provide users with valuable and meaningful pieces of textual information.

3   Linguistic Descriptions in Cognitive Cities

The purpose of building linguistic descriptions, i.e., descriptions in natural language, is to provide users with textual information which is expected to be easy to read and to understand.

Natural language is plenty of vague words (Egré and Klinedinst 2011). Therefore, dealing with vagueness in the automatic generation of human-like texts is mandatory (van Deemter 2010b). However, it is not straightforward and remains a hot open problem related to three main challenges (van Deemter 2010a): (1) Conceptualization (i.e., identifying *What to say?*), (2) Formulation (i.e., setting-up *How to say it?*), and (3) Articulation (i.e., *Verbalizing it*).

Fuzzy Logic deals naturally with vagueness and linguistic descriptions are inherent to the FST since the very beginning. Moreover, the use of interpretable fuzzy models guarantees the explanation of the behavior of the modeled systems in terms of valuable and meaningful pieces of textual information as it was thoroughly explained in the previous section. However, according to the thorough review of the state-of-the-art about text generation provided by Ramos-Soto et al. (2016b), the pioneer works in the field of natural language generation (NLG) from non-linguistic (i.e., numerical and symbolic) data (also known as data-to-text, D2T in short) were published in the 1980s out of the fuzzy community. Moreover, the pipeline proposed by Reiter and Dale (2000) has inspired many NLG/D2T applications.

In the context of the FST, Yager first introduced the linguistic summarization of data in the 1980s (Yager 1982). Then, Zadeh introduced the paradigm of computing with words and perceptions in the 1990s (Zadeh 1999). Later, Trivino and Sugeno introduced the so-called linguistic description of complex phenomena (LDCP) (Trivino and Sugeno 2013). LDCP is aimed at automatic generation of human-like textual descriptions about complex phenomena such as those taking place normally in cities. It follows a human-centric design methodology which has interpretable fuzzy models in the core. This technology has already been applied, e.g., to report about traffic information (Alvarez-Alvarez et al. 2012), about physical activity self-tracking (Sanchez-Valdes et al. 2016), or about household energy consumption (Conde-Clemente et al. 2016).

Only recently, both research communities, namely NLG/D2T and FST, have become to converge (Kacprzyk and Zadrozny 2010; Ramos-Soto et al. 2016a). Figure 1 depicts the LDCP architecture for describing complex phenomena in Cognitive Cities. It is noteworthy this is an enhanced version of the initial architecture proposed by Trivino and Sugeno (2013). We have added some of the most important elements described by Reiter and Dale (2000) with the aim of generating even more natural texts.

The LDCP design methodology is summarized as follows:

1. **Requirements Analysis Stage**. Designers must collect a corpus of natural language expressions typically considered in the application domain. They also must set the application parameters. To do so, they carefully analyze all the input elements depicted in Fig. 1:

   - *Communicative Goal*: Identifying the topics of interest for the expected audience (what to communicate?).
   - *User Model*: Identifying background and expectations of the target audience (who is the receiver?).
   - *Knowledge Source*: Identifying where the data are coming from (which is the data emitter?).
   - *Discourse History*: Tracking the temporal series of data and communication messages but also identifying the most used meaningful referring expressions (what has already been communicated? and how has it been communicated?).

2. **Off-line Modeling Stage**. Firstly, designers must build an interpretable fuzzy system (IFS) ready to interpret the given data related to the phenomenon under study. They follow the guidelines described in Sect. 2.2 with the aim of maximizing interpretability. The result is a granular linguistic model of the considered phenomenon (GLMP) understandable to humans. Notice that IFS/GLMP is an interpretable GLMP, i.e., a GLMP aware of fuzzy systems interpretability issues. It is a general-purpose hierarchical model with different levels of granularity.

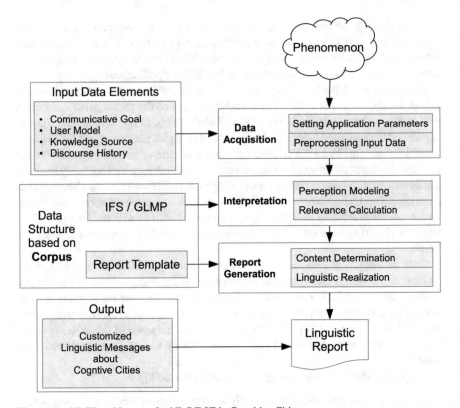

Fig. 1 The LDCP architecture for NLG/D2T in Cognitive Cities

Moreover, it is implemented as a hierarchy of Computational Perceptions (CPs) and Perception Mappings (PMs).

A CP is a tuple (Z, W_Z, R_Z) which describes a unit of information or granule in the sense of Zadeh, i.e., a clump of elements which are drawn together by indistinguishability, similarity, proximity or functionality (Zadeh 1996):

$Z = (z_1, z_2, \cdots, z_n)$ is a vector of Z-numbers (Yager 2012; Zadeh 2011) which describes the user's perception regarding a real-valued variable X. For example, let X be *"the velocity of a bus"*, then $z_1 = (high, moderate)$ means *velocity* is perceived as *high* with *moderate* reliability.

Formally, $z_i = (a_j, b_k)$ where:

- $a_j \in A = (a_1, a_2, \cdots, a_n)$, being A a vector of n everyday use linguistic expressions about the monitored phenomenon. Moreover, a_j corresponds to the most suitable linguistic term to describe the CP for a specific situation and granularity degree, in accordance with the background of the target audience.
- $b_k \in B = (b_1, b_2, .., b_m)$, being B a vector of m linguistic expressions which rate the reliability of this CP.

Notice that for maximizing interpretability both n and m must be kept between 2 and 7 ± 2 (Miller 1956). Moreover, $\forall\, j \in [1, n]$ and $\forall\, k \in [1, m]$ both a_j and b_k must be meaningful linguistic terms to the target audience.

$W_Z = (W_A, W_B)$ where:

$W_A = (w_{a_1}, w_{a_2}, \cdots, w_{a_n})$ is a vector of validity degrees $w_{a_i} \in [0, 1]$, with w_{a_i} associated to a_i and $\sum w_{a_i} = 1$.

$W_B = (w_{b_1}, w_{b_2}, .., w_{b_m})$ is a vector of validity degrees $w_{b_i} \in [0, 1]$, with w_{b_i} associated to b_i and $\sum w_{b_i} = 1$.

$R_Z = (r_{z_1}, r_{z_2}, \cdots, r_{z_n})$ is a vector of relevance degrees $r_{z_i} \in [0, 1]$, with r_{z_i} associated to z_i.

A PM is a tuple (U, y, g, T) which can create or aggregate CPs:

$U = (u_1, u_2, ..., u_p)$ is a vector of p input CPs where $u_i = (Z_{u_i}, W_{Z_{u_i}}, R_{Z_{u_i}})$. Notice that in case of first order perception mappings (1PMs, i.e., PMs at the bottom of the hierarchy), each u_i is just a numerical value taken from either sensors or databases.

$y = (Z_y, W_{Z_y}, R_{Z_y})$ is the output CP.

g is the processing function in charge of creating or aggregating CPs. It returns the validity (W_{Z_y}) and relevance (R_{Z_y}) degrees, being:

$$W_{Z_y} = g_{W_Z}(W_{Z_{u_1}}, W_{Z_{u_2}}, \ldots, W_{Z_{u_p}})$$
$$R_{Z_y} = g_{R_Z}(R_{Z_1}, R_{Z_2}, \ldots, R_{Z_p})$$

In the simplest case, A and B are defined by fuzzy partitions and accordingly g_{W_Z} and g_{R_Z} just compute the corresponding membership degrees μ_A and μ_B. Otherwise, any fuzzy aggregation function can be considered, e.g., fuzzy rules, quantified methods, or owa operators. It is important to remark that selected functions must be meaningful for the monitored phenomenon and respect the interpretability constraints described in the previous section.

T is a text generation algorithm ready to produce all possible sentences turning up as combinations of the linguistic expressions in A_y.

In addition, designers must define the so-called Report Template. This is a dynamic template which turns out as the result of combining all T in the GLMP with programming code.

3. **On-line Natural Language Generation Stage**. At run-time, the Data Acquisition module receives and pre-processes the input data which characterize the observed phenomenon. Pre-processing is required to adapt the input data to the format recognized by the Interpretation module. Then, data are interpreted using the previously generated IFS/GLMP. The result is a set of linguistic expressions that are valid (with different activation and relevance degrees) to describe the input data. Finally, the Report Generation module generates a linguistic report using the Report Template and the selected linguistic expressions. The final report is customized in accordance with the target audience.

Building the IFS/GLMP is a matter of careful design. For a suitable support in this task, designers can apply the Highly Interpretable Linguistic Knowledge (HILK) methodology (Alonso and Magdalena 2011b). HILK is a fuzzy modeling methodology that was conceived for carefully integrating expert and induced knowledge under the FST formalism, producing compact and robust models easily comprehensible by humans. It guides the user through a step-by-step procedure in the generation of all the elements involved in a fuzzy knowledge base, starting with the design of fuzzy partitions, then the rule-based learning and ending up with a knowledge base improvement stage which iteratively refines both partitions and rules. In summary, designing an IFS/GLMP involves:

- Characterization of CPs as linguistic variables with a small odd number of linguistic terms. This fits with the limited processing capability of humans. Each linguistic variable is characterized by a strong fuzzy partition in its UoD. This kind of partitions satisfies most constraints (e.g., coverage, distinguishability or overlapping) demanded to have interpretable partitions. Notice that increasing the granularity of the underlying fuzzy partitions produces an undesired increase in the number of linguistic expressions given to the user. This point is very important because the IFS/GLMP should contain only the strictly necessary and sufficient information to interpret properly the phenomenon under consideration.
- Once global semantics is defined, then we can define linguistic rules in the form "IF premise THEN conclusion". Both premise and conclusion are made up of linguistic propositions where the previously defined linguistic terms are assigned to the selected variables. Moreover, HILK provides designers with powerful tools for expert rule design, automatic rule learning, consistency analysis and optimization.

Finally, let us remark that NLG/D2T evaluation (Bugarin et al. 2015) is a hot research line which involves empirical tests regarding both intrinsic and extrinsic properties of the generated texts. They are expected to pay attention to the Gricean maxims (Dale and Reiter 1995; Grice 1975): (1) Quantity, i.e., "Make your contribution as informative as required for the current purposes of the exchange. Do not make your contribution more informative than is required"; (2) Quality, i.e., "Do not say what you believe to be false. Do not say that for which you lack adequate evidence"; (3) Relation, i.e., "Be relevant"; and (4) Manner, i.e.,"Avoid the obscurity of expressions, avoid ambiguity, be brief (avoid unnecessary prolixity), be orderly".

4   Use Case

A public local transport providing citizens with a high-quality service is more and more demanded in big cities where pollution is a matter of main concern. Citizens are aware that the use of public transport instead of private one is expected to reduce both acoustic and environmental pollution rates. However, the use of private transport is still dominant in most cities where citizens feel the public transport is not good enough yet. Measuring service quality is not straightforward. It depends on many factors such as availability, accessibility, comfort or punctuality.

For illustrative purpose, we will focus on the public local transport of a medium-size city. Namely, we will analyze the bus service in Gijón, Spain, with a population of about 300 k and 16 bus lines.[1] Notice that politicians have made a great effort in recent years to make Gijón become a modern "smart city".[2] Accordingly, the city council has promoted the engagement of citizens with the management systems through the digital transformation of the city. Moreover, the local government pushes for the creation of public services which follow the open and standard protocols of the so-called Internet of Things, in accordance with a comprehensive and open linked data strategy. For example, the website of the city council offers, among others, lots of open linked data related to energy, transport and pollution. These data are freely available in real-time through web services, mobile apps, etc. However, it is time to pass from a Smart to a Cognitive City, i.e., to make these data even more accessible and understandable to humans. It is time to develop human-centric intelligent applications which effectively exploit the available data for the benefit of citizens. Moreover, such applications must be not only user-friendly but also ready to assist citizens with valuable insights for decision-making.

The rest of the section shows a use case related to the bus service in Gijón. Namely, we analyze the raw data related to bus punctuality between two consecutive bus stops in one of the bus lines of the city. Moreover, the aim is to study how to translate raw data into valuable and understandable information and how to effectively convey such information to citizens.

To start, we summarize below how we applied the three stages of the LDCP design methodology in the context of the use case under study. It is noteworthy that the whole architecture of the system was implemented with the software rLDCP (Conde-Clemente et al. 2017b). Moreover, the IFS/GLMP was first designed with the software GUAJE (Alonso and Magdalena 2011a; Pancho et al. 2013), which implements the HILK methodology, and then exported to the XML format recognized by rLDCP.

4.1 Requirements Analysis Stage

- *Communicative Goal*: The aim is to provide citizens with insights about the punctuality of the bus service.
- *User Model*: The target audience comprises all kind of citizens. For the sake of brevity, we do not distinguish different citizen profiles in this study.
- *Knowledge Source*: We take as input the raw data provided by the city council as open linked data through the related web services. Once we select a specific bus stop, the web services return, among other data, the list of all buses passing by

---

[1] Gijón Public Transport Website at (http://en.bus.gijon.es/).

[2] Further details about how Gijón is becoming a smart city (in Spanish) at (http://smart.gijon.es/).

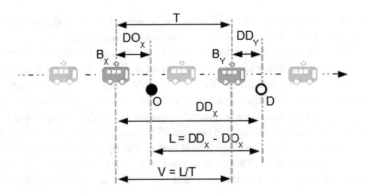

Fig. 2 Procedure to compute the estimated velocity

such stop, the estimated distance of each bus to the stop, and the last update time for the given information.

- *Discourse History*: We consider a series of data related to one week. However, for simplicity, we do not analyze the discourse history in our use case.

4.2 Off-Line Modeling Stage

The available data allow estimating the punctuality of buses with a degree of reliability. It is noteworthy that buses send periodically information related to their velocity and GPS location. However, there is no synchronization between the delivery of such information, the updating of the information provided by the web service, and the exact moment when the bus arrives to one stop.

We defined bus punctuality between two bus stops as the capability of a bus to go from the origin (O) to the destination (D) in the expected time, thus fulfilling the schedule no matter the traffic and weather conditions. Figure 2 shows graphically how we compute the estimated velocity (V) of a bus going from O to D. From all periodically reported data we consider only the two measures taken in the closest point to the next two stops, i.e., the distance to the two target stops just before arriving to them (DO_X represents the distance to the Origin while DD_X is the distance to the Destination).

In addition, with the aim of estimating the average time required to go from the origin to the destination, we selected two consecutive bus stops ("Pedro Duro" and "Playa de poniente") with six different bus lines passing by. In a preliminary experiment, we stored data for one week (seven days) and we observed how the relation between the estimated duration (T in minutes) and velocity (V in km/h) follows a hyperbolic pattern.

Then, we designed the IFS/GLMP depicted in Fig. 3. It comprises 5 PMs organized in three hierarchical levels. At the bottom, we have three numerical inputs

Fig. 3 IFS/GLMP for the
use case on studying
punctuality of buses in Gijón

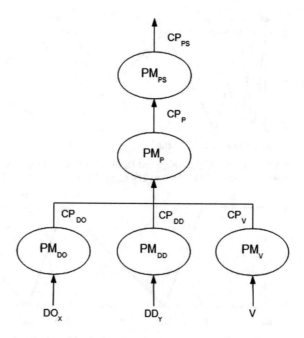

(DO_X, DD_Y, and V) which were computed as previously described (see Fig. 2). The distances to origin (DO_X) and destination (DD_Y) are measured in meters. The estimated velocity (V) is measured in km/h. These 3 inputs are processed by 3 PMs whose g are implemented in the form of interpretable fuzzy partitions (see Fig. 4). These partitions were designed with the software GUAJE which helped us to combine our expert knowledge with the knowledge extracted form the data distribution observed in the preliminary experiment. As a result, modal points in Fig. 4a reproduce the hyperbolic pattern previously observed: Extremely Small (0), Very Small (L/64), Small (L/32), Medium (L/16), Large (L/8), Very Large (L/4), Extremely Large (L/2), being L the distance between Origin and Destination, that is L = 332.8 m in our example. On the other hand, the estimated velocity (Fig. 4b) is characterized by 5 linguistic terms which correspond to the usual velocities allowed in cities (e.g., the optimal velocity is V = 30 km/h).

The text generation algorithms in the first level of the hierarchy are implemented as simple templates:

- T_{DO} = "The distance to the origin is {*Extremely Small* | *Very Small* | *Small* | *Medium* | *Large* | *Very Large* | *Extremely Large*}."
- T_{DD} = "The distance to the destination is {*Extremely Small* | *Very Small* | *Small* | *Medium* | *Large* | *Very Large* | *Extremely Large*}."
- T_V = "The estimated velocity is {*Low* | *Typical* | *Optimal* | *High* | *Very High*}."

The output of the first level of the hierarchy yields three CPs (CP_{DO}, CP_{DD} and CP_V). For simplicity, we assume all data sources have high reliability. Accordingly,

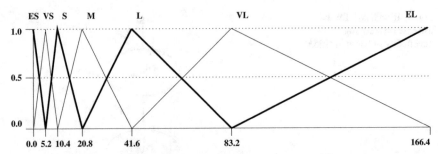

(a) Distance to Origin (DO_X) and Destination (DD_Y); being ES = Extremely Small, VS = Very Small, S = Small, M = Medium, L = Large, VL = Very Large, and EL = Extremely Large.

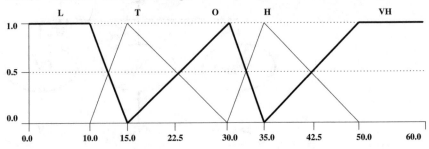

(b) Estimated Velocity (V); being L = Low, T = Typical, O = Optimal, H = high, and VH = Very High. .

Fig. 4 Strong fuzzy partitions for characterizing DO_X, DD_Y and V

all the three CPs have $B = \{High\}$. In addition, we consider all sources equally relevant. Thus, R_Z is a vector will all $r_{zi} = 1$. Notice that in more complex applications (e.g., Conde-Clemente et al. 2017a), both B and R_Z take other values and they contribute to the generation of customized texts for specific situations and for specific citizen profiles.

In the second level of the hierarchy, $PM_P = (U_P, y_P, g_P, T_P)$ characterizes the bus punctuality in the target route:

- $U_P = (CP_{DO}, CP_{DD}, CP_V)$
- $y_P = CP_P = (Z = (A, B), W_Z = (W_A, W_B), R_Z)$

  $A = \{Extremely\ Low, Very\ Low, Low, High, Very\ High\}$ describes the punctuality of the bus service.

  $B = \{Extremely\ Low, Very\ Low, Low, Medium, High, Very\ High, Extremely\ High\}$ describes the reliability of A.

- g_P is implemented by means of a fuzzy rule base which is made up of a set of Z-rules, as follows:

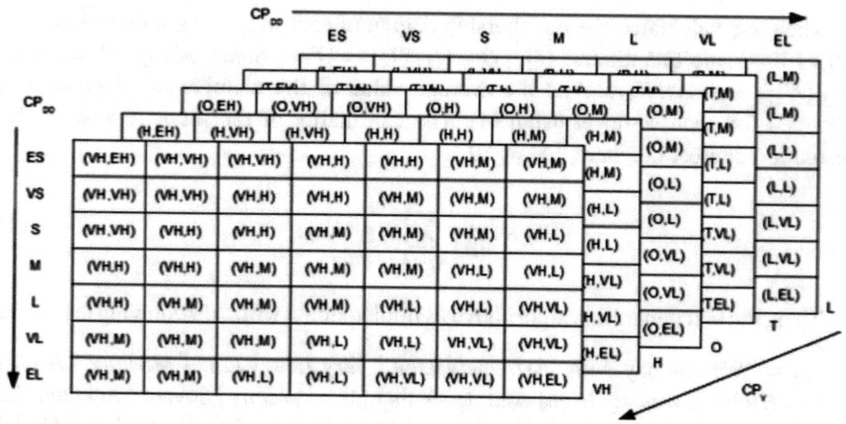

CP_{DO}	ES	VS	S	M	L	VL	EL
ES	(VH,EH)	(VH,VH)	(VH,VH)	(VH,H)	(VH,H)	(VH,M)	(VH,M)
VS	(VH,VH)	(VH,VH)	(VH,H)	(VH,H)	(VH,M)	(VH,M)	(VH,M)
S	(VH,VH)	(VH,H)	(VH,H)	(VH,M)	(VH,M)	(VH,M)	(VH,L)
M	(VH,H)	(VH,H)	(VH,M)	(VH,M)	(VH,M)	(VH,L)	(VH,L)
L	(VH,H)	(VH,M)	(VH,M)	(VH,M)	(VH,L)	(VH,L)	(VH,VL)
VL	(VH,M)	(VH,M)	(VH,M)	(VH,L)	(VH,L)	VH,VL)	(VH,VL)
EL	(VH,M)	(VH,M)	(VH,L)	(VH,L)	(VH,VL)	(VH,VL)	(VH,EL)

Fig. 5 Z-rules for computing CP_P

R_1 : **IF** CP_{DO} is (*Small, H*) AND CP_{DD} is (*Small, H*) AND CP_V is (*Optimal, H*)
 THEN CP_P is (*Optimal, High*)

R_2 : **IF** CP_{DO} is (*High, H*) AND CP_{DD} is (*Medium, H*) AND CP_V is (*Low, H*)
 THEN CP_P is (*Low, Medium*)

Notice that in our example, the rule base is made up of 245 expert rules which are sketched in Fig. 5. It is noteworthy that, as far as we know, there is not yet any learning method for extracting Z-rules from data. This task remains as a challenging future work which is beyond the scope of this book chapter.

- T_P = "We can say with {*Extremely Low | Very Low | Low | Medium | High | Very High | Extremely High*} confidence that the punctuality of the bus service is {*Extremely Low | Very Low | Low | High | Very High*}."

At the top of the hierarchy, PM_{PS} aggregates a series of CP_P. To do so, g_{PS} is an aggregation function based on the α-cuts quantified method proposed by Delgado et al. (2000). We calculated the percentage of days contained at each α−level ($N_{\alpha j}$) by (2), with $\alpha \in A = (0.1, 0.2, 0.3, 0.4, 0.5, 0.6, 0.7, 0.8, 0.9, 1)$:

$$N_{\alpha j} = \frac{1}{K} \sum_{k=1}^{K} F_\alpha(w_{1j}[k]) \tag{2}$$

where $F_\alpha(z)$ is defined by (3), and $w_{1j}[k]$ is the validity degree of CP_P, which is calculated for each day (k) being K the total number of days ($K = 7$ for one week).

$$F_\alpha(z) = \begin{cases} 1 \text{ if } z \geq \alpha \\ 0 \text{ if } z < \alpha \end{cases} \tag{3}$$

Then, we calculated the membership degree of each $N_{\alpha j}$ to each element of the set of linguistic quantifiers: $\{Q_1, Q_2, Q_3, Q_4\} = \{Few, Some, Many, Most\}$, e.g., $\mu_{Q_4}(N_{\alpha j}) = most(N_{\alpha j})$. And the average value of the membership degrees was obtained for each $\alpha-$level using (4). The cardinality of the set A represents the resolution degree, i.e., here, $|A| = 10$.

$$w_{ij} = \frac{1}{|A|} \sum_{\forall \alpha \in A} \mu_{Q_i}(N_{\alpha j}) \tag{4}$$

Then, the text generation algorithm was implemented with the following template:

- T_{PS} = "We can say with {*Extremely Low* | *Very Low* | *Low* | *Medium* | *High* | *Very High* | *Extremely High*} confidence that {*Most* | *Many* | *Some* | *Few*} days the punctuality of the bus service has been {*Extremely Low* | *Very Low* | *Low* | *High* | *Very High*}."

Finally, the Report Template was defined as follows:

"From {Day_1} to {Day_2}, {$Month$} {$Year$}, we monitored the service quality of the bus {bus_{ID}} in the line {$line_{ID}$}, from {$Origin$} to {$Destination$}. {T_{PS}}."
IF (*Exception*) **THEN** "Except for {Day_i} when {T_P}."

Exception means that one of the days under consideration the behavior of the bus was opposite to the usual one. Of course, this Report Template may be customized in accordance with the preferences of the target audience.

4.3 On-Line Natural Language Generation Stage

Let us show here a couple of examples of the reports generated at run-time by the LDCP system described in the previous section. Firstly, a case where no exception was reported:

> From Wednesday 8 to Friday 10, February 2017, we monitored the service quality of the bus 335 in the line 12, from "Pedro Duro" to "Playa de Poniente". We can say with high confidence that most days the punctuality of the bus service has been very high.

Secondly, if we slightly enlarge the selected time window, then we can find out a case where exception arises:

> From Monday 6 to Sunday 12, February 2017, we monitored the service quality of the bus 335 in the line 12, from "Pedro Duro" to"Playa de Poniente". We

can say with high confidence that most days the punctuality of the bus service has been very high. Except for Tuesday 7 when we can say with very high confidence that the punctuality of the bus service has been very low.

Probably, the very low punctuality reported on Tuesday 7 was due to some road maintenance work which took place in the morning and affected part of the selected route, thus producing heavy traffic and penalizing punctuality.

Notice that assessing the value of the derived natural language reports for the citizens remains as future work. We have just presented an illustrative use case and thus running NLG intrinsic and extrinsic evaluation tests is beyond the scope of this book chapter.

5 Conclusions and Future Work

This paper has discussed how interpretable fuzzy systems can contribute to create Cognitive Cities where the interaction between humans and machines is a major concern. Intelligent systems are expected to assist citizens in their daily life. Being recognized as universal approximators, fuzzy systems are ready to deal properly with imprecision and uncertainty when assisting citizens to solve real-world problems where applying approximate reasoning is the best choice. Moreover, interpretability is recognized as one of the main advantages of fuzzy systems since it facilitates the interaction with citizens who only can trust on systems whose behavior is understandable. Unfortunately, fuzzy systems are not interpretable per se. On the contrary, building interpretable fuzzy systems is a matter of careful design.

In this work, we have first reviewed the fundamentals of interpretable fuzzy systems modeling. Then, we have thoroughly explained how to integrate this kind of systems in the core of a methodology aimed at automatically generating reports to humans. Then, we have shown an illustrative example of how to use such methodology in the context of Cognitive Cities.

Even though we focused on a very simple example just for illustrative purposes, the general framework presented in this paper is ready to be applied to more complex use cases. For example, we plan to apply this methodology with the aim of generating reports not only for the final end-users but also for assisting fuzzy designers during modeling, test, validation and maintenance stages.

Acknowledgements This work is supported by RYC-2016-19802 (Ramón y Cajal contract), and two MINECO projects TIN2017-84796-C2-1-R (BIGBISC) and TIN2014-56633-C3-3-R (ABS4SOW). All of them funded by the Spanish "Ministerio de Economía y Competitividad". Financial support from the Xunta de Galicia (Centro singular de investigación de Galicia accreditation 2016–2019) and the European Union (European Regional Development Fund—ERDF), is gratefully acknowledged.

References

Al-Hmouz R, Pedrycz W, Balamash A (2015) Description and prediction of time series: a general framework of granular computing. Expert Syst Appl 42(10):4830–4839

Alonso JM, Magdalena L (2011a) Generating understandable and accurate fuzzy rule-based systems in a java environment. In: Fanelli A, Pedrycz W, Petrosino A (eds) Lecture notes in artificial intelligence, LNAI6857. Springer, Berlin, pp 212–219 (ISSN: 0302-9743)

Alonso JM, Magdalena L (2011b) HILK++: an interpretability-guided fuzzy modeling methodology for learning readable and comprehensible fuzzy rule-based classifiers. Soft Comput 15(10):1959–1980 (A Fusion of Foundations, Methodologies and Applications)

Alonso JM, Castiello C, Lucarelli M, Mencar C (2012) Modeling interpretable fuzzy rule-based classifiers for medical decision support. In: Medical applications of intelligent data analysis: research avancements, IGI GLOBAL, pp 255–272

Alonso JM, Castiello C, Mencar C (2015) Interpretability of fuzzy systems: current research trends and prospects. In: Kacprzyk J, Pedrycz W (eds) Springer handbook of computational intelligence. Springer, Berlin, pp 219–237. https://doi.org/10.1007/978-3-662-43505-2

Alvarez-Alvarez A, Sanchez-Valdes D, Trivino G, Sanchez A, Suarez PD (2012) Automatic linguistic report of traffic evolution in roads. Expert Syst Appl 39(12):11, 293–11, 302

Bugarin A, Marin N, Sanchez D, Trivino G (2015) Aspects of quality evaluation in linguistic descriptions of data. In: Proceedings of the IEEE international conference on fuzzy systems (FUZZ-IEEE), Istanbul, Turkey, pp 1–8

Conde-Clemente P, Alonso JM, Nunes E, Sanchez A, Trivino G (2017a) New types of computational perceptions: linguistic descriptions in deforestation analysis. Expert Syst Appl 85:46–60

Conde-Clemente P, Alonso JM, Trivino G (2017b) rLDCP: R package for text generation from data. In: Proceedings of the IEEE international conference on fuzzy systems (FUZZ-IEEE), Naples, Italy

Conde-Clemente P, Alonso JM, Trivino G (2016) Towards automatic generation of linguistic advice for saving energy at home. Soft Comput. https://doi.org/10.1007/s00500-016-2430-5

Dale R, Reiter E (1995) Computational interpretations of the Gricean maxims in the generation of referring expressions. Cogn. Sci. 19(2):233–263

Delgado M, Sanchez D, Vila MA (2000) Fuzzy cardinality based evaluation of quantified sentences. Int J Approx Reason 23:23–66

Domingos P (2012) A few useful things to know about machine learning. Commun ACM 55(10):78. https://doi.org/10.1145/2347736.2347755

Egré P, Klinedinst N (2011) Vagueness and language use. Palgrave Macmillan

Epstein SL (2015) Wanted: collaborative intelligence. Artif Intell 221:36–45. https://doi.org/10.1016/j.artint.2014.12.006

Finger M, Portmann E (2016) What are cognitive cities? In: Portmann E, Finger M (eds) Towards cognitive cities, studies in systems, decision and control, vol 63. Springer International Publishing, pp 1–11. https://doi.org/10.1007/978-3-319-33798-2_1

Grice HP (1975) Logic and conversation. In: Syntax and semantics: vol 3: speech acts. Academic Press, pp 41–58

Gunning D (2016) Explainable artificial intelligence (XAI). Technical report, Defense Advanced Research Projects Agency (DARPA), Arlington, USA, DARPA-BAA-16-53

Howard G (1983) Frames of mind: the theory of multiple intelligences. Basics, NY

Kacprzyk J, Zadrozny S (2010) Computing with words is an implementable paradigm: fuzzy queries, linguistic data summaries and natural language generation. IEEE Trans Fuzzy Syst 18(3):461–472

Mencar C, Castiello C, Cannone R, Fanelli AM (2011) Interpretability assessment of fuzzy knowledge bases: a cointension based approach. Int J Approx Reason 52(4):501–518

Miller GA (1956) The magical number seven, plus or minus two: some limits on our capacity for processing information. Psychol Rev 63(2):81–97

Mortier R, Haddadi H, Henderson T, McAuley D, Crowcroft J (2014) Human-data interaction: the human face of the data-driven society. Available at SSRN 2508051

Mostashari A, Arnold F, Mansouri M, Finger M (2011) Cognitive cities and intelligent urban governance. Netw Ind Q 13(3):4–7

Novak M (1997) Cognitive Cities. In: Intelligent environments. Elsevier, pp 386–420, https://doi.org/10.1016/B978-044482332-8/50023-8

Pancho DP, Alonso JM, Magdalena L (2013) Quest for interpretability-accuracy trade-off supported by fingrams into the fuzzy modeling tool GUAJE. Int J Comput Intell Syst 6(sup1):46–60

Pedrycz W (2012) Fuzzy modelling: paradigms and practice, vol 7. Springer Science & Business Media

Pedrycz W, Skowron A, Kreinovich V (2008) Handbook of granular computing. Wiley. https://doi.org/10.1002/9780470724163

Perticone V, Tabacchi ME (2016) Towards the improvement of citizen communication through computational intelligence. In: Portmann E, Finger M (eds) Towards cognitive cities. Springer International Publishing, pp 83–100. https://doi.org/10.1007/978-3-319-33798-2_5

Ramos-Soto A, Bugarin A, Barro S (2016a) Fuzzy sets across the natural language generation pipeline. Prog Artif Intell 5(4):261–276. https://doi.org/10.1007/s13748-016-0097-x

Ramos-Soto A, Bugarin A, Barro S (2016b) On the role of linguistic descriptions of data in the building of natural language generation systems. Fuzzy Sets Syst 285:31–51

Reiter E, Dale R (2000) Building natural language generation systems. Cambridge University Press

Russell S, Norvig P (2003) Artificial intelligence: a modern approach, 2nd edn. Prentice Hall

Sanchez-Valdes D, Alvarez-Alvarez A, Trivino G (2016) Dynamic linguistic descriptions of time series applied to self-track the physical activity. Fuzzy Sets Syst 285:162–181

Schacter D, Gilbert D, Wegner D, Hood B (2015) Psychology: second European edition. Palgrave Macmillan

Stone P, Brooks R, Brynjolfsson E, Calo R, Etzioni O, Hager G, Hirschberg J, Kalyanakrishnan S, Kamar E, Kraus S, Leyton KB, Parkes D, Press W, Saxenian A, Shah J, Tambe M, Astro T (2016) Artificial intelligence and life in 2030. Technical report, One hundred year study on artificial intelligence: report of the 2015–2016 Study Panel, Stanford University, Stanford, CA

Trivino G, Sugeno M (2013) Towards linguistic descriptions of phenomena. Int J Approx Reason 54:22–34

Tusnovics DA (2007) Cognitive Cities: interdisciplinary approach reconsidering the process of (re)inventing urban habitat. In: REAL CORP, vol 8, pp 755–764

United Nations DoE, Social Affairs PD (2014) World urbanization prospects: the 2014 revision, highlights. Technical report, (ST/ESA/SER.A/352)

van Deemter K (2010a) Computational models of referring: a study in cognitive science. The MIT Press

van Deemter K (2010b) Not exactly: in praise of vagueness. Oxford University Press

Wilke G, Portmann E (2016) Granular computing as a basis of humandata interaction: a cognitive cities use case. Granul Comput 1(3):181–197. https://doi.org/10.1007/s41066-016-0015-4

Yager RR (1982) A new approach to the summarization of data. Inf Sci 28:69–86

Yager RR (1995) Fuzzy summaries in database mining. In: IEEE conference on artificial intelligence for applications, pp 265–269

Yager RR (2012) On a view of zadeh's z-numbers. In: Proceedings of the IPMU conference, vol CCIS299. Springer, Berlin, pp 90–101

Yager RR, Zadeh LA (2012) An introduction to fuzzy logic applications in intelligent systems. The springer international series in engineering and computer science. Springer, US

Zadeh LA (1965) Fuzzy sets. Inf Control 8(3):338–353. https://doi.org/10.1016/S0019-9958(65)90241-X

Zadeh LA (1996) Fuzzy sets and information granularity. In: Klir GJ, Yuan B (eds) Fuzzy sets, fuzzy logic, and fuzzy systems. Selected papers by Lotfi A. Zadeh. World Scientific Publishing Co., Inc., pp 433–448

Zadeh LA (1999) From computing with numbers to computing with words-from manipulation of measurements to manipulation of perceptions. IEEE Trans Circuits Syst I Fundam Theory Appl 46(1):105–119. https://doi.org/10.1109/81.739259

Zadeh LA (2001) A new direction in AI: toward a computational theory of perceptions. Artif Intell Mag 22(1):73–84

Zadeh LA (2002) Toward a perception-based theory of probabilistic reasoning with imprecise probabilities. J Stat Plan Inference 105:233–264

Zadeh LA (2008) Is there a need for fuzzy logic? Inf Sci 178(13):2751–2779. https://doi.org/10.1016/j.ins.2008.02.012

Zadeh LA (2011) A note on Z-numbers. Inf Sci 181(14):2923–2932

José M. Alonso holds a M.S. (2003) and Ph.D. (2007) degree in Telecommunication Engineering from the Technical University of Madrid (UPM), Spain. He was postdoctoral researcher at the European Centre for Soft Computing (ECSC) (2007–2012), "Juan de la Cierva" postdoctoral researcher in the Department of Electronics at the University of Alcala (UAH) (2012) and at the Research Centre in Information Technologies (CiTIUS) of the University of Santiago de Compostela (USC) (2016–2018). He was Deputy Principal Researcher in the "Computing with Perceptions" Research Unit at ECSC (2012–2016). Currently, he is "Ramón y Cajal" researcher in the "Intelligent Systems" Research Group of the CiTIUS-USC, secretary of the European Society for Fuzzy Logic and Technology (EUSFLAT), Vice-chair of the Task Force on "Fuzzy Systems Software" in the Fuzzy Systems Technical Committee of the IEEE Computational Intelligence Society, and Associate Editor of the IEEE Computational Intelligence Magazine. He has been Honorary Research Fellow at the University of Aberdeen, Scotland, (2016) and Research Fellow in the Università degli Studi di Bari Aldo Moro, Italy (2017). He has published more than 100 papers in international journals, conferences and book chapters. His main research interests include explainable artificial intelligence, cognitive cities, data science, soft computing, computational intelligence, fuzzy modeling, interpretable fuzzy systems, natural language generation, free software tools, robotics and WiFi localization.

Ciro Castiello graduated (2001) and received his Ph.D. in Informatics (2005). Currently, he is an Assistant Professor at the Department of Informatics of the University of Bari Aldo Moro, Italy. His research interests include soft computing techniques, inductive learning mechanisms, interpretability of fuzzy systems, explainable artificial intelligence. He participated in several research projects and published more than seventy peer-reviewed papers. He is also regularly involved in the teaching activities of his department. He is member of the European Society for Fuzzy Logic and Technology (EUSFLAT) and of the INdAM Research Group GNCS (Italian National Group of Scientific Computing).

Corrado Mencar is Assistant Professor in Informatics at the Department of Informatics, University of Bari, Italy (since 2005). His main research interests lie in the area of computational intelligence and granular computing. In particular, he has been working for more than 15 years in the field of interpretable fuzzy systems and, more recently, in explainable artificial intelligence. He contributed to several research projects and published more than 90 peer-reviewed international papers in conference proceedings, book chapters and high-rank journals. Further-more, he teaches both undergraduate and graduate classes with chairs in programming fundamentals, information theory and granular computing courses. He is member of the European Society for Fuzzy Logic and Technology (EUSFLAT) and of the INdAM Research group GNCS (Italian National Group of Scientific Computing).

Part III
Use Cases

Towards Cognitive Cities in the Energy Domain

Javier Cuenca, Felix Larrinaga, Luka Eciolaza and Edward Curry

Abstract Current cities address efficiency challenges for optimizing the use of limited resources. City sustainability and resilience must also be improved through new learning and cognitive technologies that change citizen behavioural patterns and react to disruptive changes. These technologies will allow the evolution of current cities towards the so called "Cognitive Cities". This chapter highlights the importance of Semantic Web and semantic ontologies as a foundation for learning and cognitive systems. Energy is one of the city domains where learning and cognitive systems are needed. This chapter reviews Information and Communication Technologies (ICT)-based energy management solutions developed to improve city energy efficiency, sustainability and resilience. The review focuses on learning and cognitive solutions that improve energy sustainability and resilience through Semantic Web technologies. In addition, these solutions are evaluated from level of acceptance and use of semantics perspectives. The evaluation highlights that the Cognitive City approach is in the early stages in the energy domain and demonstrates the need for a standard energy ontology.

Keywords Cognitive cities · Energy domain · Semantic web
Semantic ontology · Energy sustainability · Energy resilience

J. Cuenca (✉) · F. Larrinaga · L. Eciolaza
Faculty of Engineering, Mondragon University, Loramendi 4,
20500 Arrasate-Mondragon, Spain
e-mail: jcuenca@mondragon.edu

F. Larrinaga
e-mail: flarrinaga@mondragon.edu

L. Eciolaza
e-mail: leciolaza@mondragon.edu

E. Curry
Insight Centre for Data Analytics, National University of Ireland Galway,
Galway, Ireland
e-mail: edward.curry@insight-centre.org

© Springer Nature Switzerland AG 2019
E. Portmann et al. (eds.), *Designing Cognitive Cities*, Studies in Systems,
Decision and Control 176, https://doi.org/10.1007/978-3-030-00317-3_7

155

1   Introduction

Cities are complex socio-technical systems that are on the edge of chaos. The amount of different actors and domains involved in the normal performance of a city, together with the challenge of responding to exponentially growing demands (energy, water, transportation, etc.) with limited resources, need sophisticated solutions that go beyond existing technological developments and innovations. Cities represent an ecosystem where the relationships between its different parts give rise to collective behaviours (Caragliu et al. 2011; Batty et al. 2012). In such a scenario, any uncertainty may produce rapidly escalating and compounding errors in the prediction of the system's future behaviour. Different actors are constantly changing their inner properties to better fit in the current environment, thus the analysis of the relations and interactions between them represents a non-trivial challenge that needs global/intersectorial solutions.

In this sense, we fully agree with the view in Finger and Portmann (2016), that *"urban problems cannot be reduced only to efficiency problems"* addressed by smart cities. In the scenario where technology, institutions and organizations co-evolve, the need of learning and cognitive technologies in order to address sustainability and resilience challenges are clear. The constant dynamic interplay between order and disorder need creative systems solutions. Cognitive Cities will address the current urban challenges of efficiency, sustainability and resilience. Efficiency refers to optimizing the use of limited resources; sustainability is about increasing humans' ecology awareness, while resilience implies the successful adaptation to changes. As defined by Finger and Portmann (2016), *"cognitive cities build on learning cities, which in turn build on smart cities"*.

In order to improve cities sustainability and resilience, future cognitive solutions for cities must make data available to different city actors and must detect and react to external shocks (i.e., economic crisis, epidemics, heat waves, water shortages, etc.). Another key feature of cognitive systems is human involvement. Human-machine interactions are needed as human-machine collaboration allows reacting to disruptive situations (Finger and Portmann 2016). The Semantic Web (Berners-Lee et al. 2001) enables all these capabilities. Semantic Web provides tools for relating and making inferences from large amounts of data from different domains. Semantic Web also provides standardized machine-readable vocabularies for data exchange and common vocabularies for human-machine interaction. Hence, we argue that Semantic Web must be the base of future cities' cognitive solutions.

In this chapter, we will focus on the situation of the energy domain within cities. Cities account for around 70% of global energy consumption and over 70% of energy-related carbon emissions (Field et al. 2014). The integration of renewable energy sources (RESs) as distributed generation is an attractive solution to deal with the dependency on fossil fuels, the constant increment of the energy consumption and the poor energy quality supplied by a conservative and aged power network. This distributed/decentralized solution represents an enhanced complexity for the management of city energy systems. Meeting the requirement for enhanced outcomes in terms of quality of life on the one hand and greater resilience (successful adaptation to fast and slow moving shocks and stressors) on the other, needs greater sophistication of governance (Moyser and Uffer 2016). While the advancement in technology has been scaled up to support cities (e.g., sensor embedded energy grids), there is still very limited demonstration of integrated information communication systems across city departments and between stakeholders. City energy management is a potential niche of application of smart, learning and cognitive systems. The purpose is to improve the current grid in terms of efficiency, sustainability and resilience to create a future Smart Grid (Fang et al. 2012).

This chapter provides a review and an evaluation on developed Information and Communication Technologies (ICT)-based solutions for improving cities energy management within recent research projects and initiatives. In the review, solutions are classified by the energy aspect (efficiency, sustainability, or resilience) on which they impact. The greatest part of the review focuses on semantic models for representing energy data and ontology-based learning and cognitive systems for improving energy sustainability and resilience. These solutions are evaluated from two perspectives: level of acceptance and use of semantics. The first perspective expresses the level of acceptance of reviewed solutions from the research and market perspectives. The second perspective identifies how the use of Semantic Web can be improved in order to accelerate the adoption of energy management cognitive solutions in future cities. Finally, the chapter enumerates short and long-term steps for achieving mass market deployment of ICT solutions towards Cognitive Cities in the energy domain.

The structure of the chapter is as follows: Sect. 2 explains the future Cognitive Cities vision and required layers. Section 3 highlights Semantic Web as the base for future cities' learning and cognitive solutions. Section 4 introduces the Cognitive Cities energy scope and provides a literature review about developed ICT-based solutions within this scope. Section 5 provides an evaluation of energy management ICT-solutions. Finally, in Sect. 6 conclusions and the future work in cities energy management solutions are presented.

2   Cognitive Cities Overview

Smart Cities integrate ICT in order to improve efficiency, addressing environmental, economic and social issues. Within the Smart Cities approach, ICT gather and analyse urban infrastructures data, such as energy, traffic, public safety or water. The purpose is to optimize these infrastructures, making possible an efficient use of usually limited resources and easing or simplifying life for citizens. On top of that, cities have to address sustainability and resilience challenges. These challenges can not only be tackled with technical solutions, but also human involvement and the ability to deal with disruptive changes (Finger and Portmann 2016). Thus, city actors must change their working habits, social relations and consumption patterns in order to improve cities economic, social and ecological sustainability. Both citizens and organizations should be provided with access to urban data analysis results so that they can learn from this information. The knowledge obtained in this stage will be used to change the city actors' behaviour. This stage is known as *Learning Cities*. City resilience requires taking a step further as technologies and actors must collaborate in order to withstand disruptive changes and external shocks (i.e., economic crisis, water/energy shortages, transport breakdowns, etc.); this requires what is known as *Cognitive Cities*.

Efficiency can be addressed by managing data generated by current ICT through smart systems that optimize the use of city infrastructures. However, sustainability and resilience require both technological improvements and human involvement (Finger and Portmann 2016). New learning and cognitive systems that change citizens' behavioural patterns and adapt to disruptive changes must be developed. Learning and cognitive systems learn about different urban environments and make decisions to improve city sustainability and resilience. This requires a set of technologies for exchanging, analysing and making inferences about data from different domains (Finger and Portmann 2016). We consider a data domain as a set of related concepts that belong to a specific area of interest (Hebeler et al. 2011). Urban data are stored at individual silos and heterogeneous devices. These factors are a very significant barrier for developing new learning and cognitive systems for cities. Semantic Web (Berners-Lee et al. 2001) is considered as the solution for overcoming interoperability problems that arise due to data heterogeneity and data silos. Semantic Web provides a set of technologies for representing, exchanging and processing data from different domains in a standardized way. Hence, semantic models for representing this data must be created as a base of Learning and Cognitive Cities.

To sum up, we can consider the paradigm of Cognitive Cities as the targeted evolution path of current cities, that will have to evolve in successive steps through Smart and Learning Cities (Finger and Portmann 2016) (see Fig. 1).

Delivering a Cognitive City requires a set of technologies that conform to the so called *Cognitive Cities' technology stack* (Finger and Portmann 2016). This stack

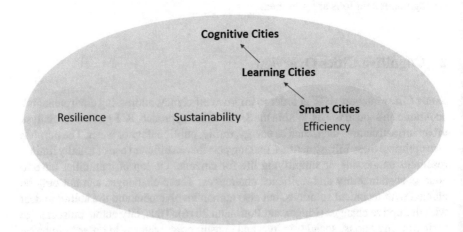

Fig. 1 Cities evolution towards Cognitive Cities

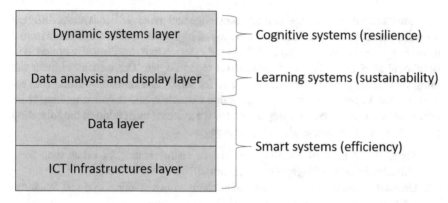

Fig. 2 Cognitive Cities' technology stack

is made up by four layers as shown in Fig. 2. Each layer is built above the previous layer and adds new technologies. All these technologies together conform *cognitive systems*.

- **ICT infrastructures layer**: this layer includes ICT-based systems that gather and exchange data from urban infrastructures using sensors and communication technologies.
- **Data layer**: this layer adds data representation standards and optimization techniques. Data representation standards are used to represent urban infrastructures data collected by the ICT infrastructure layer. Optimization techniques are used for city resources optimization purposes. Hence, standards and optimization techniques along with ICT-based systems form smart systems.
- **Data analysis and display layer**: this layer adds tools for urban infrastructure data exchange and analysis (i.e., Big Data (Cavanillas et al. 2016)). It also adds intuitive and user-centered display and social media tools for human-machine interaction. All these tools together conform *learning systems*. These systems are oriented to assist different actors on changing their behavioural patterns.
- **Dynamic layer**: this layer adds dynamic systems that detect real-time environmental changes. These systems react to environmental changes in collaboration with humans. This collaboration is enabled by new soft computing methods (i.e., natural language processing, pattern recognition algorithms, etc.) that provide a human-computer automatic interaction.

3   Semantic Web Role in Cognitive Cities

Both learning and cognitive systems must learn from different urban environments in order to assist actors in changing their behavioural patterns and adapting to disruptive changes in collaboration with humans. In other words, learning and cognitive systems are required to exchange, extract knowledge and make decisions about different domains for large volumes of data collected at high rates and in most cases in real time (Finger and Portmann 2016). ICT-based systems have traditionally operated in functional silos and rely on heterogeneous technologies. These factors hinder the integration among devices (Moyser and Uffer 2016) and human-machine interaction. Hence, in order to facilitate knowledge extraction and decision making from urban environment data, learning and cognitive systems must address the following interoperability (Serrano et al. 2015) challenges:

1. There is the need of creating a model or representation of urban data from different domains (Finger and Portmann 2016).
2. Urban data must be represented and exchanged in a standardized and machine-readable way (Moyser and Uffer 2016).

Semantic Web provides the necessary technologies for addressing these challenges. Semantic Web was defined in 2001 by Berners-Lee et al. (2001) as *"an extension of the current web in which information is given well-defined meaning, better enabling computers and people to work in cooperation."* It adds metadata to the information available on the Web, creating vocabularies that describe additional information such as the content, meaning and data relationships. This information should be meaningful and manageable by both humans and computers.

Formal representations of the Semantic Web vocabularies are called ontologies. Ontologies represent web data as a set of standard classes and objects and relations between them (Berners-Lee et al. 2001). Additionally, ontologies can include inference rules when describing and relating web data, improving intelligent agents' performance when deductions over web data. Semantic Web encompasses a set of standards and technologies used to describe and relate data on the Web. The set of best practices of using these standards and technologies is called Linked Data. Bizer et al. (2009) define Linked Data as *"data published on the Web in such a way that it is machine-readable, its meaning is explicitly defined, it is linked to other external data sets, and can in turn be linked to and from external data sets"*. Hence, the Linked Data approach allows the connection of data from different domains and data stored in different systems in order to create a global knowledge base. Future cities' learning and cognitive capabilities will benefit from Semantic Web technologies in several ways:

- Semantic Web provides standardized and machine-readable data representation, exchange and processing mechanisms for systems that rely on heterogeneous data formats as well as data access interfaces and protocols.
- As data relationships are specified, data can be linked across different domains, eliminating data silos.
- By means of using the Semantic Web humans can communicate with machines using a common vocabulary, a common set of rules and even natural language.
- Thanks to semantics, machines are capable of inferring knowledge from explicit facts.

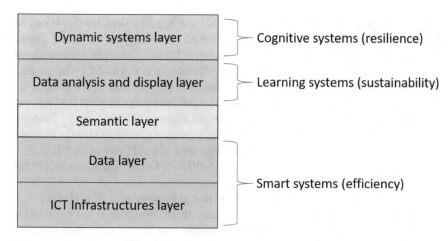

Fig. 3 Cognitive Cities' technology stack (II)

All these benefits together result in a better performance of intelligent agents and data analysis applications used for knowledge extraction and decision making within cities learning and cognitive systems. Taking this into account, the Semantic Web should be considered as an intermediate layer between the data layer and the data analysis and display layer (see Fig. 3). It provides a bridge between Smart systems and Learning systems, and by extension, cognitive systems. Semantic representation of different urban domains is a key requisite in the Smart Cities evolution process towards Cognitive Cities.

4   Cognitive Cities in the Energy Domain

Energy is one of city domains where ICT based solutions are being applied. The aim is to improve energy resources management and to integrate buildings and infras-tructures (i.e., smart homes, public buildings, organization facilities, etc.) in the future Smart Grid. According to Fang et al. (2012), the *"Smart grid is envisioned to meet the 21st century energy requirements in a sophisticated manner with real time approach by integrating ICT to the existing power grid with monitoring and control purposes"*. ICT will enable a two-way communications network between energy stakeholders, namely customers, utilities and energy operators (i.e., markets, energy service providers, Distribution System Operators (DSOs), etc.) (Locke and Gallagher 2010). This network is called Advanced Metering Infrastructure (AMI). AMIs will improve demand side management and the knowledge that energy stakeholders have about energy usage.

Furthermore, by adding ICT to the current energy grid, a scalable and reliable integration of distributed energy resources (DERs) like RESs (i.e., photovoltaics), energy storage systems (ESSs) (i.e., batteries) and Electric Vehicles (EVs) can be performed. This will lead to new scenarios such as microgrids or Virtual Power Plants (VPPs). Both microgrids and VPPs are networks that will replace conventional power plants and will improve current grid efficiency and flexibility by integrating distributed generation, ESSs and loads (Unamuno and Barrena 2015; Fang et al. 2012). Both AMIs and DERs integration require future Smart Grid applications such as, home and building energy management systems (HEMs), energy Demand Response (DR) applications, power outage management systems (OMSs), advanced power distribution management, asset management, etc. (Gungor et al. 2013). The current state of the art of these applications in different Smart Grid scenarios (Smart Homes, microgrids, etc.) is discussed in later sections.

Through these applications, Smart Grid aims to improve current grid efficiency, sustainability and resilience. Regarding efficiency, the objective is to optimize the use of both non-renewable and renewable energy sources through smart systems. Regarding sustainability, the objective is to provide citizens (i.e., energy consumers, energy auditors, building designers, etc.) a complete assessment of the energy performance of different city infrastructures (i.e., homes, public buildings, organizations), and suggest actions to change their energy management behavioural patterns for economic, social and ecological purposes through learning systems. Regarding resilience, the objective is the use of cognitive systems in collaboration with humans in order to prevent, avoid and react to power outages caused by power peak periods or natural disasters. In order to develop learning and cognitive systems within cities energy scope, energy data from different domains must be collected, exchanged, processed and analysed efficiently in real time (Rusitschka and Curry 2016). According to Corrado et al. (2015), energy data can be classified into the following categories:

- **Energy performance data**: it includes energy quantities (i.e., energy consumption, renewable energy production, etc.) energy performance indicators (i.e., CO_2 emissions, energy cost, etc.) and energy systems data (i.e., RESs, appliances, Heating, ventilation and air conditioning (HVAC) systems, etc.).
- **Energy-related and contextual data**: it includes buildings and infrastructures technical data (i.e., building construction, building geometry, etc.), geographical data (i.e., latitude and longitude, height above sea level, etc.), weather data (i.e., temperature, humidity, precipitation, etc.), environmental data (air pollutants of the urban area such as nitrogen dioxide, ozone, etc.,) socio-economic data (i.e., population income and poverty, economic activity, etc.), demographic data (i.e., population density, age, learning and education, etc.), legislative constraints (current or new infrastructures performance requirements) and land and buildings registry data (i.e., land value, land tenure, etc.).

As can be seen, these data are relevant to many domains (weather, building technical data, etc.) and are stored in heterogeneous and non-integrated devices. The ICT-based systems interoperability issues explained above are also present in the Smart Grid scope. Thus, a semantic representation of energy data is needed as an enabler/bridge to progress from Smart Grids towards learning and cognitive systems.

The following subsections provide a literature review on existing ICT-based solutions for improving cities energy efficiency, sustainability and resilience within energy management research projects and initiatives. These solutions are classified by the *Cognitive Cities' technology stack* layers on which they impact.

4.1 ICT Infrastructure Layer and Data Layer

In this layer, ICT solutions are used for data metering, data transmission, data storage and energy optimization.

Smart Grid energy data metering is performed by AMI systems, Automatic Meter Reading (AMR) systems and smart meters. These systems are used to create a two-way communication network between energy consumers and utilities. Other type of sensors and sensors networks (i.e., Phasor measurement units, Wireless Sensor Networks (WSNs)) are also used to measure and monitor Smart Grid mechanical state (Fang et al. 2012).

The energy data transmission is performed by a set of communication technologies. These technologies can be wireless or wired (Gungor et al. 2011). The main Smart Grid wireless communication technologies are the following: ZigBee, 6LoWPAN, Z-wave, Wireless Mesh and Cellular Network Communication (i.e., 3G, WiMAX) (Mahmood et al. 2015). The main Smart Grid wired communication technologies are the following: Power Line Communication (PLC) and Digital Subscriber Lines (DSL). Transmitted data are represented under a set of Smart Grid standards. These standards can be classified according to Smart Grid application where there are used (Gungor et al. 2011): Revenue metering information model (i.e., ANSI C12.19, M-Bus, etc.), building automation (i.e., BACnet), substation automation (i.e., IEC 61850), powerline networking (i.e., HomePlug, PRIME, etc.), home energy measurement and control (i.e., U-SNAP, IEEE P1901, Application-Level Energy Management Systems (i.e., IEC 61970, OpenADR), Inter-control and Inter-operability Center Communications (i.e., IEEE P2030, ANSI C12.22), Cyber Security (i.e., IEC 62351) and Electric Vehicles (i.e., SAE J2293, SAE 2836).

Smart Grid energy data can be stored in different repositories: Relational Database Management Systems (RDBMS) (i.e., Oracle, MySQL, Microsoft SQL Server, etc.), NoSQL databases (i.e., Cassandra, HBase, MongoDB, etc.) and cloud-based distributed file systems (i.e., Google File Systems, Hadoop Distributed file system, Disco Distributed File System, etc.).

Energy optimisation at this level is concerned with finding optimal solutions for contexts where maximization and minimization techniques can be applied. An example of an optimization problem can be to minimize as much as household energy consumption while maintaining a specific comfort temperature. When applying optimization techniques, the optimization problem is represented mathematically through an objective function. This function is subject to a set of constraints represented as equations or inequalities (Snyman 2005). In the previous example, the objective function would be to *minimize the energy consumption while the constraint would be maintaining a specific comfort temperature.*

In the energy scope, optimization techniques have been applied with different ecological, environmental, and operational objectives: maximize the revenue, minimize carbon emissions, maximize reliability, maximize energy production, minimize operation cost, minimize investment cost, etc. Optimization techniques allow optimizing some of these aspects at a time and taking into account several constraints (a minimum energy production, a maximum operation cost, etc.). Iqbal et al. (2014) and Fathima and Palanisamy (2015) provide insights about the application of different optimization techniques for improving Smart Grid energy efficiency. These optimization techniques are classified into two main categories: Linear optimization techniques or Linear Programming (LP) and nonlinear optimization techniques or Nonlinear Programming (NLP). LP is applied when the optimization problem is represented as a linear function and constraints (linear optimization problems). NLP is applied when the optimization problem is represented as a non-linear function and constraints (nonlinear optimization problems) (Luenberger et al. 1984). Linear and non-linear optimization techniques applied for improving energy efficiency include: Simplex algorithm (Luenberger et al. 1984), Nelder-Mead algorithm (Singer and Nelder 2009), meta-heuristics (Glover and Kochenberger 2006) and Artificial Neural Networks (ANNs) (Fathima and Palanisamy 2015). Apart from previous algorithms, there are also sets of optimization tools that have been used for improving Smart Grid energy efficiency. The most commonly used tools are HOMER (Lambert et al. 2005), GAMS (Fathima and Palanisamy 2015) and HYBRID2 (Baring-Gould et al. 1996).

4.2 Semantic Layer

The concern of this review in the semantic layer is to identify the different ontologies used to represent the energy domain. From the beginning of the current decade, Semantic Web technologies were applied for creating ontologies that represent energy data of different Smart Grid scenarios such as Smart Homes, buildings, organizations or microgrids. These ontologies are aimed to be the energy knowledge base for Smart Grid applications that are in the conceptual or design phases.

On the one hand, Kofler et al. (2012) and Daniele et al. (2016) present Smart Homes energy data representation models. Kofler et al. (2012) present an ontology design created within the ThinkHome project.[1] The ontology represents, in a machine-readable way, home energy consumption, production and energy-related contextual data. The ontology is made up by several ontologies that represent different domains data:

---

[1] http://www.eui.eu/Projects/THINK/Home.aspx.

- *Building ontology*: it represents building architecture data i.e., layout, spaces data, etc.
- *User information ontology*: it represents user comfort preferences, user schedules, etc.
- *Processes ontology*: it represents home system processes, user activities data, etc.
- *Exterior ontology*: it represents weather and climate conditions.
- *Energy and resource parameter ontology*: it represents home equipment or devices (i.e., home appliances, energy measurement sensors) data, home energy demand and supply data and energy providers and energy tariffs data.

Furthermore, the authors suggest how represented data can be used combined with a multi-agent system in order **to improve energy efficiency at future smart homes**. The proposed use cases are:

- Select energy providers depending on produced energy type or energy tariffs (i.e., consume only energy produced by RESs or select a provider which has an excess of energy and sells it cheaper).
- Disconnect unnecessary equipment according to occupancy or customer behaviour patterns (i.e., disconnect from the electricity grid entertainment equipment such as the TV when user is unlikely to return more to the living room).

Daniele et al. (2016) present the ontology SAREF (Daniele et al. 2015) and its current version, SAREF4EE. The objective of the ontology SAREF4EE is **to improve interoperability among electrical appliances of different manufacturers** allowing them to be connected with customer energy management systems used for Smart Grid DR optimization strategies. The SAREF4EE ontology represents the following information: home appliances, sensors and actuators data (i.e., device manufacturer, device state, device function, energy flexibility, etc.), building spaces (i.e., rooms), home energy production and consumption data, associated costs, energy performance data time intervals and home weather conditions and home occupancy data.

On the other hand, Blomqvist et al. (2014) and Andreas Fernbach and Kastner (2015) present building energy data representation models. Blomqvist et al. (2014) publishes as published as Linked Data the data about energy efficiency improvements, energy saving recommendations and energy measures taken from previous energy audits are within DEFRAM and DEFRAM-2 projects.[2] The linked dataset published represents the following data: energy audits and measures of industrial organizations and recommendations for improving energy management given after previous audits. It also represents data about investment cost of applying such recommendations, achieved energy saves and additional information about the organization (i.e., organization location, organization facility size, etc.). The final purpose is to use the previous linked dataset as a knowledge base for future ICT-based solutions **to help organizations for saving energy based on energy audits performed over similar organizations, to facilitate researches and policy makers comparing and analysing data from different audits and to facilitate third parties' applications that use energy audits data**.

---

[2]http://www.ida.liu.se/~evabl45/defram.en.shtml.

Andreas Fernbach and Kastner (2015) present an ontology that describes building features and Building Automation Systems (BASs). BASs monitor and control automatically HVAC systems of indoor environments (Kastner et al. 2005). The ontology is presented as a first step of using Semantic Web technologies for the **automated integration of BASs developed by different manufacturers**, and represents the following information: static building information (i.e., architectural, geometrical, building topology, building physics properties, etc.) and building equipment technical information (HVAC systems as well as lighting applications). The ontology also represents BASs configuration data such as device (i.e., measurement equipment, HVAC systems controllers and actuators, etc.) locations, functionalities and datapoint descriptions.

The ontology OntoMG (Salameh et al. 2015) represents a microgrid energy data. A microgrid is a set of RESs, ESSs and loads that can operate autonomously or connected to the main grid. The ontology OntoMG is presented as the knowledge base of a microgrid energy management system that is being developed. The ontology encompasses renewable and non-renewable generators, storage equipment, electrically connected loads and their properties, which include mobility, economical, operational and ecological aspects. The purpose of this ontology is to be used by computational and optimization techniques aiming **to achieve different microgrid objectives (i.e., minimizing transmission losses, generating good power quality, minimization of green-house effect gases, etc.)**.

Finally, Hippolyte et al. (2016) and Gillani et al. (2014) present energy data representation models for Smart Grid wider areas. Hippolyte et al. (2016) provide a general approach of Semantic Web application for representing Smart Grid prosumers energy data within the MAS2TERING European project.[3] A prosumer is a Smart Grid stakeholder that consumes and produces energy. Specifically, the ontology MAS2TERING is aimed **to facilitate the representation the data of different Smart Grid domains and provide interoperability among different Smart Grid agents and stakeholders**. The MAS2TERING ontology links concepts of data representation standards used in different energy domains. These concepts are the following: home area networks (smart appliances, power profiles, renewable energy generation, smart meters and smart user interfaces), energy DR concepts (i.e., market context, dynamic pricing and event descriptions, etc.) and Smart Grid stakeholders' information and their roles and responsibilities within both the energy supply value chain and the flexibility value chain. The authors' final purpose is to use this ontology as a base for Smart Grid multi-agent systems for an energy market coordination process for improving energy flexibility among energy prosumers and DSOs.

Gillani et al. (2014) present an ontology for representing energy data of prosumer oriented Smart Grids. The ontology is aimed to be complemented with an inductive reasoning layer. This layer is in the design phase and will contain applications for detecting the energy consumption patterns of consumer appliances, energy production patterns and energy producers' performance (i.e., efficiency, impact to the environment) patterns. The objective is **to improve Smart Grid DR and sustainability**

---

[3]http://www.mas2tering.eu/.

by predicting Smart Grid energy consumption and production. The prosumer oriented Smart Grid ontology represents the following data: infrastructures data (type of operation, time and geographical location, and power critical premises), electrical appliances data (consumption and temporal data, power consumption rating and operational patterns), electrical generation systems data, power storage systems data (type, produced power, charge and discharge efficiency, etc.), weather report data, events (i.e., electrical appliance events, weather events, storage events and generator events, etc.), energy production and consumption services contractual information and connectivity relationships between producers and consumers.

4.3 Data Analysis and Display Layer

The review in this layer presents ICT solutions that go one step further enabling the construction of learning systems focused on improving Smart Grid sustainability. These systems use different data analysis techniques and display tools over semantically represented energy data models. Learning systems provide citizens a holistic view of infrastructures energy performance and suggest actions for changing their energy management behavioural patterns. With these systems home energy consumers and both public and private organizations will see their energy bills slashed. They also will be able to choose between a wide variety of energy vendors depending on their energy tariffs. Public and private organizations will also be benefited, as they will perform a more efficiency management of their energy consumption sources (i.e., facilities, business travel, etc.) with both economic and ecological purposes (Curry et al. 2012).

On the one hand, the solutions proposed by Curry et al. (2012), Hu et al. (2016), Niknam and Karshenas (2015) and Pont et al. (2015) are oriented to assess citizens about urban infrastructures energy performance. Curry et al. (2012) present an enterprise energy observatory system. The aim of this system is **to improve enterprise energy management at different levels from both economic and ecological perspectives**. The enterprise energy observatory system includes data analysis and display applications that provide an enterprise energy performance view at organizational, function and individual level:

- *Organizational level*: executives can view the real-time consumption of energy across all enterprises domains, IT, facilities, travel, etc.
- *Function level*: the system provides a fine-grained understanding of what business activities are responsible for IT energy usage, and can enable IT to bill appropriately.
- *Individual level*: it gives an employee real-time energy consumption data on their IT, Facilities, Travel, etc.

The system includes also internal applications (i.e., a Complex Event Processing (CEP) engine, data search and query engines, etc.) that ease the knowledge extraction of enterprise Linked Data by energy analysis applications. All system applications are underpinned by energy related data from different enterprise domains that have been published as Linked Data. This data includes enterprise business entities (i.e., employees, products, customers, equipment, assets, buildings, rooms, etc.), direct electricity consumed by Office IT and Data centers, energy consumption measurement sensors and business information (i.e., enterprise Resource Planning (ERP), finance, facility management, human resources, asset management and code compliance, etc.).

Hu et al. (2016) present a building Energy Performance Assessment (EPA) system developed within the SuperB project.[4] This system **shows the performance gap between building predicted and measured energy performance data**. The EPA system includes tools that measure, analyse and show building or particular zones energy performance data. The energy performance data are expressed as energy metrics that include Energy Use Intensity (EUI), energy cost, normalised atmospheric emissions, etc. These metrics are compared with building predicted energy performance data. A building energy performance simulation model makes these predictions. Data used by the EPA system analysis and display tools is represented under an ontology (Corry et al. 2015) that contains and links/fuses building data of different domains. Each domain is represented by an individual ontology:

- *IfcOWL ontology*: it includes building geometry data, material properties, as-built construction details and HVAC systems specifications.
- *SIMModel ontology*: it includes building performance simulation data.
- *SSN ontology*: it contains building sensors data (i.e., consumption metering sensors, temperature sensors).
- *Performance assessment ontology*: it contains building energy performance quantitative metrics needed to compare current with predicted energy performance.

Sensor measurements values and corresponding time intervals are stored in relational databases due to performance reasons and a mapper module is used to link previous ontologies with measurement and time values.

Niknam and Karshenas (2015) and Pont et al. (2015) also present building EPA systems, but in this case these systems are focused on the design stage. The EPA system developed by (Niknam and Karshenas 2015) **shows building designers the building energy performance corresponding to a building specific design**. The objective is to optimize the building design for a better energy performance. Specifically, a prototype of the EPA system was developed that predicts building heating cost based on its design and simulated environmental conditions data through a heating cost calculation algorithm. The EPA system is underpinned by four ontologies that represent the following data respectively: building properties (i.e., surface area, thickness, heat transfer coefficient, etc.), mechanical equipment specifications (i.e., capacity, type of fuel, and energy consumption, etc.), historical weather information of building geographic location and energy cost information based on the type of energy required for mechanical equipment.

---

[4]http://cordis.europa.eu/project/rcn/187015_en.html.

Pont et al. (2015) present a web decision support and optimization platform for building designers. The purpose of the web platform is to make buildings energy performance-oriented designs within the SEMERGY project.[5] This platform **shows building designers' suggestions about different building components alternatives according to user preferences and technical constraints for optimizing heating demand, environmental impact and investment cost**. These suggestions are made by a reasoning interface that makes inferences through building design and simulated environmental conditions data. These data are represented by an ontology that captures building geometry and material data, building equipment data, building materials data and historical and simulated weather data of building geographic location.

An integrated platform[6] that shows energy related data about cities to different actors is presented within the SEMANCO project.[7] The aim of this platform is to **provide a complete view of city energy performance in order to help different city actors** (i.e., energy policy makers, building designers, citizens, etc.) **to make informed decisions for reducing cities carbon emissions**. The platform includes visualization tools that display energy data and analysis tools that perform different analysis tasks (i.e., make energy performance predictions, classify buildings according to their consumption or carbon emissions, etc.) over cities energy data at different scales (building, neighbourhood, municipality or region). The integrated platform is underpinned by an ontology that captures energy efficiency concepts of urban areas (Corrado et al. 2015). The objective of this ontology is to provide models for urban energy systems to be able **to assess the energy performance of an urban area**. The ontology represents the following information: building energy consumption data, associated energy performance indicators (i.e., energy savings, energy costs, etc.) and timestamps, consumed energy sources, building features, building equipment features and services. The ontology also represents external factors such as weather conditions, building geographical location, demographic, environmental and socio-economic data.

The solutions presented by Burel et al. (2016), Fensel et al. (2014), Sicilia et al. (2015), Yuce and Rezgui (2015) and Stavropoulos et al. (2016), apart from offering energy assessment, are oriented **to offer citizens suggestions for improving urban infrastructures energy performance**. Burel et al. (2016) present the EnergyUse collaborative web platform. The purpose of this platform is **to raise home end-user climate change awareness**. The platform collects home appliances energy consumption data from smart plugs and allows end users viewing and comparing the actual energy consumption of various appliances. Users can also share energy consumption values with other users and create open discussions about energy saving tips. Discussions are described and classified by tags defined by users. These tags correspond to energy appliances and topics related with the discussed energy saving tips. The EnergyUse platform includes tools that analyse and extract concepts

---

[5]http://www.semergy.net/.

[6]http://www.semanco-project.eu/index_htm_files/SEMANCO_D5.4_20131028.pdf.

[7]http://semanco-project.eu/.

from discussions created. These tools link extracted concepts with appliance and environmental terms included in external semantic repositories in order to create new tags and descriptions for discussions. The purpose of these additional tags and descriptions is to improve user navigation experience among discussions. Finally, the EnergyUse platform also exports appliance consumption and community generated energy tips as linked data to be used by third parties, such as other users or websites. The EnergyUse platform is supported by the ontology EnergyUse, which represents the following information: user profiles of users that use the platform, home appliances and HVAC systems data, home sensors and actuators data, home appliances energy consumption measures and energy tips discussion data.

Fensel et al. (2014) present a home energy management platform developed within SESAME and SESAME-S[8] projects. The aim of this platform is **to help home users making better decisions in order to reduce their energy consumption**. The platform allows users defining energy saving policies and it generates its own energy saving policies through an ontology reasoning engine. Specifically, this ontology reasoning engine generates schedules and rules for turning on and off home devices based on tariff plans and desired indoor environmental conditions. Energy saving policies are presented through different user interfaces aimed to stimulate and facilitate users to use energy more responsibly. Home energy data are represented under the following ontologies (Fensel et al. 2013):

- *SESAME Automation Ontology*: it represents general concepts (e.g. resident data, location data) and home automation and energy domain data (i.e., device, configuration).
- *SESAME Meter Data Ontology*: it represents metering equipment data.
- *SESAME Pricing Ontology*: it represents available energy types and tariffs data.

Sicilia et al. (2015) present a web-based Decision Support System (DSS) prototype developed within the framework of the OPTIMUS project.[9] The objective of the DSS is **to support users and organizations decision-making process for improving buildings energy efficiency**. The DSS uses machine learning algorithms to predict building energy performance and environmental conditions. Predictions take into account seven different energy data domains. These domains are represented and linked by the ontology OPTIMUS: building/equipment features data, weather forecasting data, energy and environmental values measured by sensors, building occupants notion of comfort, building occupants comfort patterns, energy prices data and renewable energy production data.

Yuce and Rezgui (2015) present a building energy management system that **assists users to save energy** developed within the KnoholEM Project.[10] This system is underpinned by a semantic knowledge database which contains building information and devices metering data. These data are used by an ANN that learns building

---

[8]http://sesame-s.ftw.at/.

[9]http://www.optimus-smartcity.eu/.

[10]http://www.knoholem.eu/page.jsp?id=2.

consumption patterns, and a genetic algorithm (GA)-based optimization tool that generates optimized energy saving rules taking into account learned energy consumption patterns and different objectives (including comfort) and constraints. These rules are presented to facility managers as energy saving suggestions through a graphical user interface (GUI).

Stavropoulos et al. (2016) present a building energy management system that combines energy assessment, energy advice and building automation. This system **monitors building energy performance and shows this information to allow users taking actions to increment energy savings.** Intelligent agents within the system also devise short-term and long-term energy saving policies that are automatically generated and enforced. Furthermore, the system is also designed to receive energy providers' instructions in future Smart Grids. This system is supported by the ontology BOnSAI (Stavropoulos et al. 2012), which represents the following energy data: building appliances and sensor/actuators data, building structure data, user location and energy and environmental condition measures.

4.4 Dynamic Layer

Finally, the ICT solutions identified in the review and related to the dynamic layer concentrate on improving Smart Grid resilience. Specifically, these solutions are focused on improving Smart Grid DR by detecting disruptive situations (i.e., power peak periods) over semantically modelled data. These systems react to disruptive situations in collaboration with humans.

Zhou et al. (2012b) present a CEP engine (Zhou et al. 2012c) developed within the Los Angeles Smart Grid Demonstration Project.[11] The CEP engine purpose is **to enable dynamic DR applications that detect power peak situations and perform actions to improve DR.** The CEP engine is supported by a Smart Grid semantic information model (Zhou et al. 2012a) that is made up of different ontologies in order to represent different energy data domains:

- *Electrical equipment ontology*: it contains electrical equipment features and power consumption details collected from smart meters.
- *Organizations ontology*: it contains different organizations information, people involved in the organization as well, as their roles within the organization.
- *Infrastructures ontology*: it contains environment concepts including transportation networks, buildings and so on, besides the Power Grid infrastructure.
- *Weather ontology*: it contains weather information.
- *Spatial ontology* it contains building equipment or infrastructure spatial location information.
- *Temporal ontology*: it contains power consumption time data, scheduling information of infrastructure, electrical equipment data and individual people data.

---

[11] https://www.smartgrid.gov/project/los_angeles_department_water_and_power_smart_grid_regional_demonstration.html.

Shi et al. (2014) present a **microgrid energy management and control system that combines both sustainability and resilience actions is presented**. Hence, this system impacts on both data analysis and display and dynamic layers. On the one hand, the microgrid energy management system includes a Human Machine Interface (HMI) for microgrid monitoring and control. Apart from that, the system includes a microgrid scheduling algorithm and a microgrid DR optimization algorithm. The DR optimization algorithm adapts microgrid demand to real-time energy prices. The energy-scheduling algorithm schedules microgrid DERs and loads with both economic and ecological optimization purposes. Both algorithms use semantically represented data that includes: microgrid devices information (i.e., DERs, smart meters, smart appliances, EVs charging station, PVs and batteries, etc.), weather forecast information, Automated Demand Response (ADR) signals received from utility and energy market information.

Finally, Zhang et al. (2016) present an energy management platform for VPPs. VPPs are groups of DERs and controllable loads that act as a single energy stakeholder within the Smart Grid. Within VPPs energy prosumers sell their surplus energy during energy curtailment or energy consumption peak load periods. The energy management platform adapts VPPs energy production and consumption to peak loads that occur both either in the VPP or the Smart Grid. The energy management platform includes algorithms that **select the best energy storage systems scheduling strategy among energy prosumers for facing energy peak load periods in Smart Grid and VPP in a distributed manner**. The selection of the strategy is based on energy generation sources and loads, respective energy generation and consumption forecasting performed by machine learning algorithms (i.e., Dynamic Bayesian Networks). All information used by the platform to manage VPPs energy DR is represented by an ontology. This information includes: buildings and facilities data, buildings spatial-use patterns, energy production, consumption and storage systems (i.e., renewable energy generation units and controllable loads) data, ICT based sensors and actuators data. The ontology also represents weather conditions of areas where systems are deployed, events (i.e., prosumer energy consumption or generation changes, weather condition changes and loads operation changes, etc.) and services offered by prosumers for improving VPP or Smart Grid DR such as energy supply or energy curtailment.

5 Evaluation

This section presents the evaluation of the literature survey in relation to semantics and the advances towards Cognitive Cities for the energy domain. These solutions are evaluated from two perspectives: level of acceptance and use of semantics.

5.1 Level of Acceptance

This perspective expresses the level of acceptance of reviewed solutions from the research and market perspectives. According to Curry et al. (2016), city ICT-based solutions development are divided into two cycles. These solutions includes solutions developed for the energy domain. The first cycle corresponds to a research phase that includes experimental design and pilot deployment. The second cycle is focused on citywide deployments of ICT solutions to drive mass market adoption. Curry et al. (2016) also point out that Smart City ICT based solutions have reached this second cycle, as current Smart City projects are focused on key innovation characteristics (i.e., relative advantage, compatibility, cost efficiency, risk level, etc.) for mass market adoption.

This cyclic approach can be also extended to learning and cognitive systems for the energy domain. Regarding Learning Cities, there is a lot of literature about ontology-based learning systems focused on changing citizens' energy management behavioural patterns. All these systems are limited to pilot demonstrators that in some cases were implemented in specific Smart Grid scenarios. For example, the EPA system developed by Hu et al. (2016) has been implemented in a sports centre in order to measure its energy operation against previously predicted performance by a BEPS model; the enterprise energy observatory system has been implemented within the Irish Insight Centre for Data Analytics (formerly DERI: Digital Enterprise Research Institute); the DSS developed by Sicilia et al. (2015) has been validated in three different buildings in order to predict their energy performance; the building energy management system developed by Yuce and Rezgui (2015) has been tested in a care home; and the integrated platform developed within SEMANCO project has been tested in three cities for analysing energy performance data. The next step is to evaluate the impact, compatibility, cost efficiency, feasibility and benefits of these systems in citywide deployments (Curry et al. 2016). After evaluating these aspects, learning systems shall be marketed to consumers. There is less literature about cognitive systems focused on improving Smart Grid resilience. These solutions are still in the experimental design (Zhou et al. 2012b) or pilot demonstrator implementation phases: the microgrid energy management and control system developed by Shi et al. (2014) has been implemented in a pilot microgrid and VPP energy management platform developed by Zhang et al. (2016) has been tested in a pilot VPP.

In conclusion, we can say that the scientific community investigating ICT-based solutions is evolving towards Cognitive Cities and is in the early stages in the energy domain. From the market point of view, only Smart City initiatives are tackling ICT-based solutions innovation aspects (see Fig. 4).

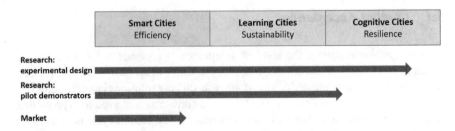

Fig. 4 Evaluation of Cognitive Cities' energy scope ICT solutions progress

5.2 *Use of Semantics*

The second perspective identifies how the use of Semantic Web can be improved in order to accelerate the adoption of energy management cognitive solutions in future cities. As said in Sects. 4.2, 4.3 and 4.4, the ontologies reviewed in this chapter are/will be the knowledge base of a wide variety of Smart Grid energy management applications. All these applications can be grouped into five high-level categories. Each category corresponds to a Smart Grid scenario on which the Smart Grid application impacts: Smart Home energy management, building/district/city energy management, organization energy management, microgrid energy management and Smart Grid DR management.

Within previous categories, each Smart Grid application belongs to a specific scope of application. For example, within Smart Home energy management, there are applications that are focused on energy assessment and device control, energy saving collaborative advice, etc. In many cases, represented energy data domains are repeated among developed ontologies. This is particularly true when ontologies are/will be the knowledge base of applications that belong to the same category. For example, most of ontologies developed within the reviewed Smart Grid applications represent the building technical equipment data. Table 1 illustrates which energy data domains have been included in ontologies according to Smart Grid application category and scope of application.

Table 1 shows that *Basic energy related concepts* are represented in most ontologies. External factors such as climate and geographical data are also present in most ontologies. We cannot say the same about other external factors such as environmental, demographic and socio-economic data. Specific equipment (Non-renewable energy sources, RESs and ESSs) data are mainly represented at microgrid and Smart Grid DR energy management applications ontologies. There are some exceptions as building and Smart Home energy management applications use RESs and ESSs data. Almost all Smart Grid stakeholders are represented in DR management applications ontologies. Other applications only use specific stakeholder data depending on their scope of application. Smart Grid DR data is limited to Smart Grid DR management applications. Energy performance data (apart from energy consumption and production) is present at building and Smart Grid DR management applications.

Table 1 Represented energy data domains depending on Smart Grid application category and scope of application

Application categories and scope of application	Smart Home energy management			Building / district / city energy management		Organization energy management			Microgrid energy management	Smart Grid DR energy management	
Energy data domains	Energy assessment and device control (Kofler et al. 2012)	Energy saving collaborative advice (Burel et al. 2016)	Energy saving advice and control (Fensel et al. 2014)	BAS integration (Andreas Fernbach and Kastner, 2015)	Building EPA (Hu et al. 2016) (Niksaan and Kamhenas, 2015) (Pont et al. 2015) (Corrado et al. 2015)	Energy saving advice (Sicilia et al. 2015) (Yuce and Rezgui, 2015) (Stavropoulos et al. 2016)	Energy saving advice (Blomqvist et al. 2014)	Energy assessment (Curry et al. 2012)	Microgrid multi-objective energy optimization (Salameh et al. 2015) (Shi et al. 2014)	Smart Grid energy market coordination process (Hippolyte et al. 2016) (Gillani et al. 2014) (Zhou et al. 2012b) (Daniele et al. 2016)	VPP energy market coordination (Zhang et al. 2016)
Basic energy related concepts Building / infrastructure technical data	x	x	x	x	x	x	x	x	·	x	x
Energy consumption systems data	x	x	x	x	x	x	x	x	x	x	x
Energy production / consumption data	x	x	x	x	x	x	x	x	x	x	x
Sensors / Actuators data	x	x	x	x	x	x	·	x	x	x	x
Specific equipment data RESs data	x	·	x	·	·	x	·	·	x	x	x
ESSs data	·	·	·	x	·	x	·	·	x	x	x
BASs devices data	·	·	·	x	·	·	·	·	·	·	·
External factors Weather / climate data	x	x	x	x	x	x	x	·	x	x	x
Geographical data	x	x	x	x	x	x	x	x	x	x	x
Environmental data	·	·	·	·	x	x	·	·	·	·	·
Demographic and socio-economic data	·	·	·	·	x	·	·	·	·	·	·
Smart Grid users / stakeholders data Home user data	x	x	x	x	·	x	·	·	·	x	·
Building occupants data	·	·	·	x	·	·	·	·	·	x	·
Employees data	·	·	·	·	·	·	·	x	·	x	·
Energy suppliers / utilities data	x	·	x	x	x	x	·	x	x	x	x
DSOs data	·	·	·	·	·	·	·	·	·	x	·
Smart Grid DR data DR operations	·	·	·	·	·	·	·	·	x	x	·
Events data	·	·	·	·	·	x	·	·	·	x	x
Services offered by prosumers	·	·	·	·	·	·	·	·	·	·	x
Energy performance data Energy Key Performance Indicators (KPIs) data	·	·	·	·	x	·	·	·	·	x	·
Economical, operational and ecological objectives	·	·	·	·	·	·	·	·	·	·	x
Building energy performance stimulation data	·	·	·	·	x	·	·	·	·	·	·
Organization-related data Organization facilities data	·	·	·	·	·	·	x	x	·	x	·
Organization assets data	·	·	·	·	·	·	·	x	·	x	·
Business processes data	·	·	·	·	·	·	·	x	·	·	·
Investment cost of energy saving plans	·	·	·	·	·	·	x	·	·	·	·
Energy audits data	·	·	·	·	·	·	x	·	·	·	·
Other concepts Home processes data	x	·	x	·	·	·	·	·	·	·	·
Energy saving tips and recommendations data	·	x	·	·	·	x	·	·	·	·	·

Organization related data is limited to organization energy management applications with the exception of Smart Grid DR management applications. Some Smart Home energy management applications include concepts such as home processes data in their ontologies. Finally, energy saving tips and recommendations are included in Smart Home and buildings energy saving applications. Table 1 also shows that applications which ontologies belong to a specific category introduce new concepts in their ontologies. For example, all ontologies of Smart Home energy management applications represent home user data. Other concepts, on the other hand, are only included in applications that belong to a specific scope of application. For example, only the energy saving collaborative advice application includes home energy tips data in its ontology.

A common data representation model is one of Smart Grid key challenges that can be addressed by Semantic Web technologies (Wagner et al. 2010). The energy data domains repetition among reviewed ontologies evidences a convergence towards a standard energy ontology. One single ontology can be used in a wide variety of Smart Grid scenarios and energy management applications with minimal changes. A standard energy ontology should include at first Basic energy related concepts enumerated in Table 1. Other energy domains that can be present in more than one Smart Grid scenarios should be also included. These energy domains are: RESs and ESSs data, external factors (i.e., weather/climate data, environmental data, etc.), Smart Grid users/stakeholders (i.e., home users, organizations, building occupants, etc.) data, Smart Grid DR operations data, energy Key Performance Indicators (KPIs) and energy saving tips and recommendations data. Then, applications that belong to a specific scope of application (i.e., home energy collaborative advice) can add application specific concepts (i.e., energy tips discussions) to the standard energy ontology. Looking further ahead, as cognitive energy management applications evolve and settle in cities, a nearly fully standard energy ontology can be developed. There will be no need to add new data as new energy management applications arise.

A standard energy ontology would also help to represent energy domains in a common manner. Energy data domains are represented with different levels of detail among reviewed ontologies. This is particularly true for *Basic energy related concepts*. Table 2 shows the level of detail some of reviewed energy ontologies represent *Basic energy related concepts*.

When considering energy consumption systems data, ThinkHome, EnergyUse and ProSGV3 ontologies are candidates to represent this domain. However, one of these ontologies may be not enough to represent the whole energy consumption systems domain. One ontology may include energy systems data that other ontology does not. For example, the EnergyUse ontology represents heating systems while BOnSAI ontology does not. Figure 5 shows which concepts include each ontology regarding energy consumption systems domain. ThinkHome, EnergyUse and

Table 2 Basic energy related concepts representation level of detail (H = High/M = Medium/L = Low)

Energy domains	Ontology					
	ThinkHome ontology[a]	DEFRAM project ontology[b]	SAREF4EE ontology[c]	BOnSAI ontology[d]	EnergyUse ontology[e]	ProSGV3 ontology[f]
Building/infrastructure technical data	H	–	L	M	H	H
Energy consumption systems data	H	–	M	L	H	H
Energy production/ consumption data	H	M	H	H	H	H
Sensors/actuators data	H	–	M	M	–	M

[a]https://www.auto.tuwien.ac.at/downloads/thinkhome/ontology/,
[b]http://www.ida.liu.se/projects/semtech/schemas/energy/2013/09/efficiency.owl,
[c]http://ontology.tno.nl/saref4ee/,
[d]http://lpis.csd.auth.gr/ontologies/bonsai/BOnSAI.owl,
[e]http://eelst.cs.unibo.it/apps/LODE/source?; http://socsem.open.ac.uk/ontologies/eu,
[f]http://data-satin.telecom-st-etienne.fr/ontologies/smartgrids/proSGV3/ProSG.html

ProSGV3 are the ontologies that include more concepts. However, each ontology (except BOnSAI) includes its own concepts. All these concepts should be merged in a standard ontology. In addition, different terms are used to represent the same energy data.

This term and domain representation diversity, called *semantic heterogeneity*, leads to an interoperability problem that hinders the full adoption of these ontologies in real scenarios (Maree and Belkhatir 2015). Hence, there is the need of creating a unified ontology that represents all energy domains providing a common terminology. This ontology could be a standard knowledge base of energy management solutions in any Smart Grid scenario or even energy management solutions applied in various scenarios at the same time, i.e., organizations that include microgrids. Moreover, a unified ontology would reduce energy management application developers' effort when creating energy ontologies and to be more focused on application implementation.

Apart from energy data ontologies standardization, Smart Grid applications should include more energy concepts that they do. Energy recommendations are only included in energy saving applications. These recommendations should be also oriented to avoid undesired situations within Smart Grid applications focused on improving grid resilience. Energy performance indicators should also be included at organizational level. Finally, in Smart Grid DR applications more stakeholders should be included; i.e., Energy Services Companies (ESCOs), Balance Responsible Parties (BRPs), etc.

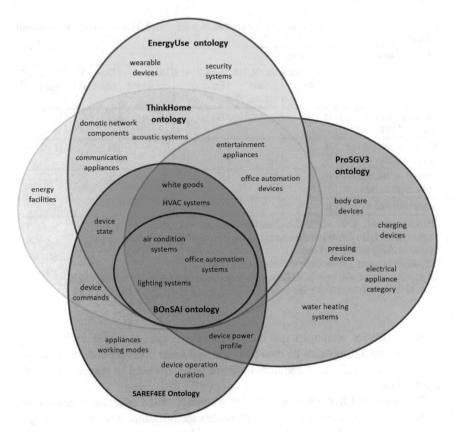

Fig. 5 Energy consumption systems data representation

6   Conclusions and Future Work

The evolution of current cities will take them to the so-called Smart Cities. Smart Cities are aimed to optimize urban infrastructures through smart systems, making possible an efficient use of the usually limited resources and a high quality of life for citizens. Apart from efficiency, cities address sustainability and resilience challenges. These aspects are addressed by learning and cognitive systems that change the citizens' behavioural patterns and adapt to disruptive changes in collaboration with humans. Cities that include learning and cognitive systems are called "Cognitive Cities". We can consider Cognitive Cities as the targeted evolution path of current cities.

Both learning and cognitive systems must learn from different urban environments in order to assist actors in changing their behavioural patterns and adapting to external shocks (i.e., economic crisis, epidemics, heat waves, water shortages, etc.) in collaboration with humans. The Semantic Web enables these capabilities. Semantic

Web provides tools for relating and making inferences from large amounts of data from different domains. Semantic Web also provides standardized machine-readable vocabularies for data exchange and common vocabularies for human-machine interaction. Hence, in this chapter we argue that Semantic Web must be the base of future cities' cognitive solutions.

City energy management is a potential niche of application of smart, learning and cognitive systems. The purpose is to improve current grid efficiency, sustainability and resilience within the future Smart Grid. In recent years, Information and Communication Technologies (ICT)-based systems applied in city energy management have evolved considerably. In this chapter, a review and evaluation of existing ICT-based solutions for improving city energy management is presented. The main part of the review is focused on sustainability and resilience aspects.

This chapter provides a review and an evaluation on developed ICT-based solutions for improving cities energy management within recent research projects and initiatives. In the review, solutions are classified by the energy aspect (efficiency, sustainability, or resilience) on which they impact. The greatest part of the review focuses on semantic models for representing energy data and ontology-based learning and cognitive systems for improving energy sustainability and resilience. Current energy efficiency solutions correspond to smart systems. These systems include energy data metering infrastructure, energy data exchange communication technologies and standards and energy data storage repositories. Then, optimization techniques are applied over energy data in order to optimize city energy management from ecological, economic and operational perspectives. Learning systems are focused on improving city energy sustainability. These systems use different data analysis techniques and user-centered display tools over energy data. Learning systems provide citizens a holistic view of infrastructures energy performance and to suggest actions for changing their energy management behavioural patterns. Energy resilience solutions correspond to cognitive systems. These systems are focused on improving Smart Grid Demand Response (DR) by detecting disruptive situations (i.e., power peak periods) and interacting with energy users.

The chapter also evaluates these solutions from two perspectives: level of acceptance and use of semantics. The first perspective expresses the level of acceptance of reviewed solutions from the research and market perspectives. According to this evaluation, development of smart systems in the energy domain is in the advanced stages as Smart Cities projects are now focusing on mass market adoption. Although there is a wide literature about learning systems in the energy domain, these systems are limited to pilot demonstrators. In some cases, the pilot demonstrators were implemented in specific Smart Grid scenarios (i.e., Smart Homes, microgrids, etc.). Before marketing energy management learning systems, they must be evaluated in citywide deployments. There is less literature about cognitive systems in the energy domain. These solutions are still in the experimental design or pilot demonstrator implementation phases.

The second perspective identifies how the use of Semantic Web can be improved in order to accelerate the adoption of energy management cognitive solutions in future cities. Semantic Web technologies application in cities energy scope within recent research initiatives has supported a breakthrough in Smart Grid energy data representation, exchange and processing. The developed ontologies within these initiatives link and represent energy performance data and energy-related data from different domains (i.e., energy performance, weather/climate data, building technical data, etc.). The representation of these energy domains depends on the Smart Grid scenario where ontologies are applied. This, along with terminological differences when representing energy data, evidences the need for a standard energy ontology. This ontology will be used in a wide variety of Smart Grid energy management applications (i.e., Smart Home energy management, microgrid energy management, etc.). In addition, a standard energy ontology will allow representing different energy domains using a common terminology.

Taking into account previous evaluations, in order to reach mass market deployment of ICT-based solutions towards Cognitive cities in the energy domain, a number of steps are necessary:

- Semantic technologies adoption in the form of ontologies as support to upper layers.
- Deployment of pilot demonstrators based on the semantic layer and providing ICT-based solutions to deal with resilience.
- Thorough test of those demonstrators.
- Include technology capabilities in commercial devices.

This vision of the future has a short-term plan, a mid-term plan and a long-term plan. The previous first two points are framed in the short term while the third and the fourth steps are vision as actions for the mid-long term.

References

Andreas Fernbach IP, Kastner W (2015) Linked data for building management

Baring-Gould EI, Green H, Van Dijk V, Manwell J (1996) Hybrid2-the hybrid power system simulation model. Technical report, USDOE, Washington, DC, USA, National Renewable Energy Laboratory, Golden, CO, USA

Batty M, Axhausen KW, Giannotti F, Pozdnoukhov A, Bazzani A, Wachowicz M, Ouzounis G, Portugali Y (2012) Smart cities of the future. Eur Phys J Special Topics 214(1):481–518

Berners-Lee T, Hendler J, Lassila O (2001) The semantic web. Sci Am 284(5):28–37

Bizer C, Heath T, Berners-Lee T (2009) Linked data-the story so far. In: Emerging concepts, semantic services, interoperability and web applications, pp 205–227

Blomqvist E, Thollanderb P, Keskisärkkä R, Paramonovab S (2014) Energy efficiency measures as linked open data

Burel G, Piccolo LS, Alani H (2016) Energyuse-a collective semantic platform for monitoring and discussing energy consumption. International semantic web conference. Springer, Heidelberg, pp 257–272

Caragliu A, Del Bo C, Nijkamp P (2011) Smart cities in Europe. J Urban Technol 18(2):65–82

Cavanillas JM, Curry E, Wahlster W (2016) New horizons for a data-driven economy

Corrado V, Ballarini I, Madrazo L, Nemirovskij G (2015) Data structuring for the ontological modelling of urban energy systems: the experience of the semanco project. Sustain Cities Soc 14:223–235

Corry E, Pauwels P, Hu S, Keane M, O'Donnell J (2015) A performance assessment ontology for the environmental and energy management of buildings. Autom Construct 57:249–259

Curry E, Hasan S, O'Riain S (2012) Enterprise energy management using a linked dataspace for energy intelligence. In: Sustainable internet and ICT for sustainability (SustainIT). IEEE, pp 1–6

Curry E, Dustdar S, Sheng QZ, Sheth A (2016) Smart cities-enabling services and applications. J Internet Serv Appl 7(1):6

Daniele L, den Hartog F, Roes J (2015) Created in close interaction with the industry: the smart appliances reference (SAREF) ontology. In: International workshop formal ontologies meet industries. Springer, pp 100–112

Daniele L, Solanki M, den Hartog F, Roes J (2016) Interoperability for smart appliances in the IoT world. International semantic web conference. Springer, Cham, pp 21–29

Fang X, Misra S, Xue G, Yang D (2012) Smart grid: the new and improved power grid: a survey. IEEE Commun Surv Tutor 14(4):944–980

Fensel A, Tomic S, Kumar V, Stefanovic M, Aleshin SV, Novikov DO (2013) SESAME-S: semantic smart home system for energy efficiency. Informatik-Spektrum 36(1):46–57

Fensel A, Kumar V, Tomic SDK (2014) End-user interfaces for energy-efficient semantically enabled smart homes. Energy Effic 7(4):655–675

Fathima AH, Palanisamy K (2015) Optimization in microgrids with hybrid energy systems-a review. Renew Sustain Energy Rev 45:431–446

Field CB, Barros VR, Mach K, Mastrandrea M (2014) Climate change 2014: impacts, adaptation, and vulnerability, vol 1. Cambridge University Press, Cambridge, New York

Finger M, Portmann E (2016) What are cognitive cities? Towards cognitive cities. Springer, Cham, pp 1–11

Gillani S, Laforest F, Picard G (2014) A generic ontology for prosumer-oriented smart grid. In: EDBT/ICDT workshops, pp 134–139

Glover FW, Kochenberger GA (2006) Handbook of metaheuristics, vol 57. Springer Science & Business Media, Boston

Gungor VC, Sahin D, Kocak T, Ergut S, Buccella C, Cecati C, Hancke GP (2011) Smart grid technologies: communication technologies and standards. IEEE Trans Ind Inf 7(4):529–539

Gungor VC, Sahin D, Kocak T, Ergut S, Buccella C, Cecati C, Hancke GP (2013) A survey on smart grid potential applications and communication requirements. IEEE Trans Ind Inf 9(1):28–42

Hebeler J, Fisher M, Blace R, Perez-Lopez A (2011) Semantic web programming. Wiley

Hippolyte J, Howell S, Yuce B, Mourshed M, Sleiman H, Vinyals M, Vanhee L (2016) Ontology-based demand-side flexibility management in smart grids using a multi-agent system. In: 2016 IEEE International smart cities conference (ISC2). IEEE, pp 1–7

Hu S, Corry E, Curry E, Turner WJ, O'Donnell J (2016) Building performance optimisation: a hybrid architecture for the integration of contextual information and time-series data. Autom Construct 70:51–61

Iqbal M, Azam M, Naeem M, Khwaja A, Anpalagan A (2014) Optimization classification, algorithms and tools for renewable energy: a review. Renew Sustain Energy Rev 39:640–654

Kastner W, Neugschwandtner G, Soucek S, Newman HM (2005) Communication systems for building automation and control. Proc IEEE 93(6):1178–1203

Kofler MJ, Reinisch C, Kastner W (2012) A semantic representation of energy-related information in future smart homes. Energy Build 47:169–179

Lambert T, Gilman P, Lilienthal P (2005) Micropower system modeling with homer. In: Farret FA, Godoy Simoes M (eds) Integration of alternative sources of energy

Locke G, Gallagher PD (2010) NIST framework and roadmap for smart grid interoperability standards, release 1.0. National Institute of Standards and Technology, vol 33

Luenberger DG, Ye Y, et al (1984) Linear and nonlinear programming, vol 2. Springer

Mahmood A, Javaid N, Razzaq S (2015) A review of wireless communications for smart grid. Renew Sustain Energy Rev 41:248–260

Maree M, Belkhatir M (2015) Addressing semantic heterogeneity through multiple knowledge base assisted merging of domain-specific ontologies. Knowl-Based Syst 73:199–211

Moyser R, Uffer S (2016) From smart to cognitive: a roadmap for the adoption of technology in cities. In: Towards cognitive cities, Springer, pp 13–35

Niknam M, Karshenas S (2015) Sustainable design of buildings using semantic BIM and semantic web services. Proc Eng 118:909–917

Pont UJ, Ghiassi N, Fenz S, Heurix J, Mahdavi A (2015) SEMERGY: application of semantic web technologies in performance-guided building design optimization. J Inf Technol Construct 20:107–120. http://www.itcon.org

Rusitschka S, Curry E (2016) Big data in the energy and transport sectors. In: New horizons for a data-driven economy. Springer, pp 225–244

Salameh K, Chbeir R, Camblong H, Tekli G, Vechiu I (2015) A generic ontology-based information model for better management of microgrids. In: IFIP international conference on artificial intelligence applications and innovations. Springer, pp 451–466

Serrano M, Barnaghi P, Carrez F, Cousin P, Vermesan O, Friess P (2015) Internet of things IoT semantic interoperability: research challenges, best practices, recommendations and next steps. In: European research cluster on the internet of things

Shi W, Lee EK, Yao D, Huang R, Chu CC, Gadh R (2014) Evaluating microgrid management and control with an implementable energy management system. In: IEEE international conference on smart grid communications (SmartGridComm). IEEE, pp 272–277

Sicilia Á, Costa G, Corrado V, Gorrino A, Corno F (2015) A semantic decision support system to optimize the energy use of public buildings

Singer S, Nelder J (2009) Nelder-mead algorithm. Scholarpedia 4(7):2928

Snyman J (2005) Practical mathematical optimization: an introduction to basic optimization theory and classical and new gradient-based algorithms, vol 97. Springer Science & Business Media

Stavropoulos TG, Vrakas D, Vlachava D, Bassiliades N (2012) Bonsai: a smart building ontology for ambient intelligence. In: Proceedings of the 2nd international conference on web intelligence, mining and semantics. ACM, p 30

Stavropoulos TG, Koutitas G, Vrakas D, Kontopoulos E, Vlahavas I (2016) A smart university platform for building energy monitoring and savings. J Ambient Intell Smart Environ 8(3):301–323

Unamuno E, Barrena JA (2015) Hybrid ac/dc microgrids part I: review and classification of topologies. Renew Sustain Energy Rev 52:1251–1259

Wagner A, Speiser S, Harth A (2010) Semantic web technologies for a smart energy grid: Requirements and challenges. In: Proceedings of the 2010 international conference on posters & demonstrations track, vol 658, CEUR-WS.org, pp 33–36

Yuce B, Rezgui Y (2015) An ANN-GA semantic rule-based system to reduce the gap between predicted and actual energy consumption in buildings

Zhang J, Seet BC, Lie TT (2016) An event-based resource management framework for distributed decision-making in decentralized virtual power plants. Energies 9(8):595

Zhou Q, Natarajan S, Simmhan Y, Prasanna V (2012a) Semantic information modeling for emerging applications in smart grid. In: 2012 Ninth international conference on information technology: new generations (ITNG). IEEE, pp 775–782

Zhou Q, Simmhan Y, Prasanna V (2012b) Incorporating semantic knowledge into dynamic data processing for smart power grids. In: International semantic web conference. Springer, pp 257–273

Zhou Q, Simmhan Y, Prasanna V (2012c) SCEPter: semantic complex event processing over end-to-end data flows. Technical Report 12-926, Computer Science Department, University of Southern California

Javier Cuenca is a Ph.D. Student at the Software and Systems Engineering research group, Mondragon University (Basque Country, Spain). He graduated in Computer Science and Engineering (2014) and in Embedded Systems (2016), both at Mondragon University. During his master studies, he worked at the Electronics and Computing Department of the Mondragon University. During this period, he worked in several European projects related with smart energy (ARROWHEAD, RENNOVATES and CITYFIED). In his thesis project he focuses on applying Semantic Web technologies and ontology engineering techniques to represent the energy data domain as a base for future energy sustainability and resilience applications. Javier Cuenca is the author of several papers published at international conferences such as IECON in Florence (2016) or WOMocOE in Vienna (2017).

Felix Larrinaga is currently lecturer and researcher at Mondragon University (Basque Country, Spain). Since 2000, he combines lecturing and research at Mondragon University with an extensive experience in industrial work as a project leader. He has actively participated in the creation of the new research team on Web Engineering at the Faculty of Engineering. The research activities of the group focuses on interoperability, social web technologies, and semantic web technologies. The team works on multidisciplinary projects with other universities, technological centres and enterprises in areas such as innovation management, health or tourism. Currently, Felix Larrinaga works on several projects related to manufacturing, smart cities and energy (Productive 4.0, MANTIS, ARROWHEAD, CITYFIED, RENNOVATES and SMARTENCITY). He obtained his PhD in Mobile Communications at Staffordshire University (UK) in 1999. Previously, he finished the BSc (Hons) in Industrial Electronics at Mondragon University (1995) and the MSc in Computing Engineering (2010). Expertise: Web engineering, interoperability, social web technologies, semantic web.

Luka Eciolaza Ph.D. is a lecturer and researcher at the Robotics and Automation research group, within the department of Electronics and Computation at Mondragon University (Basque Country, Spain). He has wide experience in the areas of data mining, advanced reasoning and control of non-linear systems. He has been involved in many international research and development projects. Currently, he takes part in activities of Intelligent Manufacturing for the analysis, control and optimization of industrial processes.

Edward Curry is a Research Leader at the Insight Centre for Data Analytics and a funded investigator at LERO, The Irish Software Research Centre. His research projects include studies of smart cities, energy intelligence, semantic information management, event-based systems, stream processing and human-in-the-loop. Edward Curry has worked extensively with industry and government advising on the adoption patterns, practicalities, and benefits of new technologies. Edward has published over 130 scientific articles in journals, books, and international conferences. He has presented at numerous events and has given invited talks at Berkeley, Stanford, and MIT. He is Vice President of the Big Data Value Association, a non-profit industry-led organization with the objective of increasing the competitiveness of European companies with data-driven innovation. Edward Curry is a member of the scientific leadership committee of Insight, and a Lecturer in Informatics at the National University of Ireland Galway (NUIG).

A Conceptual Model for Intelligent Urban Governance: Influencing Energy Behaviour in Cognitive Cities

Mo Mansouri and Nasrin Khansari

Abstract The premise of a cognitive city is that by having the right information at the right time, citizens, service providers and city government alike will be able to make better decisions that result in increased quality of life for urban residents and the overall sustainability of the city. This chapter explores the influence of governance on sustainability and maps the impact of information and communication technologies (ICTs) on the decision-making process by increasing policy effectiveness, accountability, and transparency within urban systems. This chapter further presents conceptual systems models for the cognitive city and energy behaviour including three sub-layers as human-institutional, physical, and data layers. This chapter offers these integrated conceptual models to increase the efficiency of energy systems under complex and uncertainty conditions, facilitate energy consumption problem solving, and support the development of capacities at the individual, social, and technical levels to improve managing energy consumptions in the future. Using a systems methodology, the socio-technical systems models explore how practices of information sharing and different structures of governance and management impact the willingness of individuals and households to adopt sustainable energy practices.

1 A New Paradigm of Governing

We are living in the era of data abundance. The high frequency of streaming the continuous and never-ending new collections of data is changing the previously adopted approaches and models, almost in any field of science and technology. Availability of data collected through sensing capabilities in urban systems of our time has

M. Mansouri (✉)
School of Systems and Enterprises, Stevens Institute of Technology,
Hoboken, NJ 07030, USA
e-mail: mmansour@stevens.edu

N. Khansari
Electrical & Systems Engineering Department, University of Pennsylvania,
Philadelphia, PA 19104, USA
e-mail: khansari@seas.upenn.edu

© Springer Nature Switzerland AG 2019
E. Portmann et al. (eds.), *Designing Cognitive Cities*, Studies in Systems,
Decision and Control 176, https://doi.org/10.1007/978-3-030-00317-3_8

made revolutionary changes in how cities are designed, planed and viewed. It has also brought about new paradigms to the horizon of city planning and management through applications of scientific methods and computational reasoning, capsulated and introduced as "the new science of cities" (Batty 2013). In the same manner, advancement of technology and availability of data can make a breakthrough in governing mechanisms applied to cities. The new governing mechanisms can be adopted from the researches done in systems sciences on management of complex systems or system of systems (Darabi et al. 2013; Mansouri and Mostashari 2010) and then institutionalized and applied to the cities as the networked environment of variety of urban systems, using the existing management frameworks (Darabi et al. 2012).

This necessity is partially affected by the new movement in the world toward designing, engineering and building cognitive cities in which all the stakeholders are provided with the effective options for adopting better choices. This can bring about a revolution in the way we make decisions in urban environments and indeed, involves the complexity of interactions among all social stakeholders including governing bodies, lawmakers (including and not limited to policymakers), private and non-profit sectors as well as citizens at large. The efforts for capturing these complexities, beyond any shadow of doubt, are transforming how cities are supposed to be governed in the world.

Like any other stakeholder-based approach, such new governing model requires a systemic perspective through which emerging school of thoughts in governing urban systems evolve. Considering the intensity of interactions based on the structure of power in society in general, and importance of proposing a holistic solution for the situation in front, governing approaches should consider a balanced interaction between bottom-up evolving processes and top-down organizing enforcements, in favour of effectiveness. This new paradigm brings cities to the level of an adaptive organism that is in constant interaction with its environment; the very capability that was distinguished by pioneers of city sciences (Jacobs 1961). The interactive dynamics of all actors in urban environment supposedly creates the force of change and growth, when aligned meticulously with planning of critical facilities and trans-actional movements within the urban systems as is argued in the literature (Jacobs 1969).

To understand this new outlook to governing urban systems, however, first we need to have an understanding of technological advancements and sensing capabilities. Technology has a direct impact on our capability to build cognitive cities that are capable of learning collectively throughout the time and evolve accordingly to adapt the emergence of their environment (Batty 2013). Ultimately, the next step should include discussions on new visions toward changing behavioural patterns of citizens at large. Despite their level of individual power, citizens are the main stakeholders of cities and their collective actions will equip urban systems with variety of smart solutions in future. These solutions are the key factors in keeping the balance in all aspects of life including economic and ecological sustainability as well as energy, water, transportation, communication, education, and healthcare. In better words, "smart" cities are the ones that can learn over time and possibly affect the behaviour of their citizens through structural and systemic influences. That type of learning organisms could be considered as cognitive cities in which the process of learning is established and hence, they have the capacity of adopting effective and efficient policies through intelligent governing methods (Mostashari et al. 2011). This chapter is dedicated to introduction of the concepts that give life to the idea of developing a new science for governing urban systems in particular urban energy system through applications of social sciences perspectives and soft systems methodology.

2 Political Aspects of Effective Governance

Governance could be considered the process through which the public power that is enacted by infrastructure of authority uses to coordinate activities, facilitate communications, increase interoperability, resolve conflicts, assist in policy-making, and improve effective participation (Mansouri and Mostashari 2010). The emphasis however, should be on the continually complicated relationship between the state and society that is supposedly the environment that embodies the network of stakeholders who play an essential role in policymaking. Therefore, effective governance happens where the state shifts its own role from direct policy control toward policy co-ordination (Bache and Flinders 2004).

There is variety of definitions of the term in the literature, mostly from political perspective, from implementation of public policy services (Rhodes 1997) to regulatory, processes for coordination and control (Osborne 2010) to a more general approach as the patterns through which order is implied (Bevir 2010). Some other organisations have presented characteristic-based definitions. For instance, The World Bank emphasizes on indicators such as accountability, stability and absence of terror or violence, and "good governance" (Lodge 2010).

In general, the literature has covered the concept within the following contextual areas: corporate governance, which focuses on organisational governing matters; good governance that has a political approach, mostly from The World Bank perspective (and those of other international similar institutes) and models of social, political and administrative governing directions; and public governance with a collaborative approach to the political, economic, social and administrative aspects of governance (Mansouri and Mostashari 2010). A role-based definition of the term categorizes governance according to its societal roles to: socio-political, which resolves social conflicts in society; public policy, which is concerned with the interaction among agents of society; administrative that focuses on running the processes in society; contract and transaction, which focuses on the contractual interactions within the actors of the societal system; and network governance that includes the concepts of "self-organisation" in complex networks (Osborne 2010).

Moreover, from a functional oriented definition, political science literature (Shahin and Finger 2008; Rossel and Finger 2007) puts a lot of value on transforming from a hierarchical governing structure toward a flat or horizontal one at which collaboration of all stakeholders is the key. This newer (and mostly European form of governing model) have three distinct dimensions: (1) it separates itself from the "government" by relating to non-state stakeholders (actors); (2) it gets operated through multiple political levels in entire society; and (3) it has a different way of running the functions of the state, mostly through actor-based approaches.

In this situation, the state (government) shares its power with non-state actors such as international or transnational corporations, which has stake and interest outside the jurisdictions of the state; as well as with non-governmental organisations (NGOs) or non-profit corporations and other philanthropy organisations focused on serving public on matters that are not in the government's interest. This behoves a multi-layered type of public management covering both global and national (or in some contexts: local and regional) levels. The role of government, consequently, gets limited to: making policies, delivering government driven services, and setting regulations through which laws emerge and executed (Zwahr et al. 2005).

In such circumstances, the competition will be not only among the corporations but also between the government and this politically empowered public sector, which includes all the aforementioned non-governmental (or non-state) actors. The competition however, will be on the share each of these stakeholders should have in territorial authorities such as: inclusiveness of decision rights and degrees of freedom in regulatory boundaries. At the same time, both types of stakeholders have to collaborate, cooperate, and in the most general sense, participate in activities related to making policies and providing services at national, supranational, and infra-national levels (Finger and Pécoud 2003).

The plethora of international transactions in the age of globalisation along with the advancements of civic participation through civil society including but not limited to national and transnational NGOs, non-profits, and other philanthropy organisations has contributed to an increase in the level of political participation and particularly in governing activities such as making policies. This change can be evidently observed from the increase in transnational economic activities of such organisations in comparison to other types of corporations as well as proliferation of regional, international and transnational institutions (Bevir 2010). This has also caused a paradigm shift in governing methods from vertical to horizontal. The classical (or rather traditional) approach to governance that involves hierarchical management based on a command and control mechanism, is gradually changing to one of self-organizing, and distributed over the network.

From a philosophical point of view, at least for nations that cherish more traditional ways of governing activities or for the kind of public services with a local range of importance and relevance, a local scale might be still favoured over a national or international scale. However, even within the realm of those types of environments, local associational actions are getting connected in variety of ways. This coupled with the advancement of democratic processes, strengthen the ability of decentralized governance, which provide more connectivity as well as responsive and coordinated actions when necessary (Chaskin and Abunimah 1999). Along with this dynamic, macro-social changes also contributed to emergence of globalisation and regionalisation around the world. This only intensifies complexity of societal transactions through scaling up interdependencies in all economic, social and political interactions. As a result, the governmental and market forces are no longer suitable for the emerging social conditions and new born problems (Jessop 1998).

3 Urban Governance and Governing Cognitive Cities

Cities have been becoming the source of economic growth not only because they are the centres of consumption, but also by representing a major part in economic, political, and social innovation. They are moreover, promoting and consolidating international competitiveness. These are also partially the result of the informational revolution and a consequence of technological advancement. The economy is driving on information and analytics. In fact, the information technology revolution had created a new type of society that is driven by knowledge in all aspects of life (Jessop 2002). At the same time, developing community is essentially relying on making decision in a distributed fashion and at a local level, where political, economic, environmental, and social needs are better known and their connection is visible (Green and Haines 2011).

The same way, technology and knowledge is driving cities toward a more flourishing economy through incentivizing competition, consumption and innovation, shifting toward decentralized resource and decision system strengthen social participation along with democracy that is presumably increasing governmental accountability. From this perspective, a city-centric power structure pushes societies from highly ideological to more practical politics that takes every stakeholder's interest into account. On the other hand, citizens who are among the actors or stakeholders of the society will be able to involve themselves more directly in issues related to their immediate interests.

From a theoretical standpoint too, some in the literature (Lummis 1996) believe in the importance of distribution of power and localisation in governance and its connection with democracy, reasoning that real stakeholders live in local areas and effectiveness of the governance in addressing their issues should be voice from those people directly. Based on the same perspective and according to the principles of urban governance approach, political systems of the city are supposed to overlook and lead governing mechanisms used to coordinate local states through their actions and use of resources (Pierre 2005). To achieve the goals of urban governance that is realisation of stakeholders' interest, all citizens should ideally participate in policy-making processes, particularly when it comes to public services. The role of a central government would be providing basic freedoms within the society and for all citizens; such as freedom of speech and religion. Central government is also supposed to defend property rights, protect free markets, and maintain social stability for their citizens according to law (Pierre 2005).

Ultimately, the real governance structure will come to action through the interactions among many actors as stakeholders of the society. The city and citizens from one side will have to collaborate with urban and national governmental departments and agencies, corporations in local, national and international levels, political parties and other political powers in the society, NGOs, activist agencies, entrepreneurs, philanthropists and non-profits, community-action groups, religious groups, etc. (Healey 2004). With such consideration, civil society actors will gain a paramount role in urban governance (Pierre 1999) and consequently, the political interactions among civil society and other sectors of the society such as government and private (corporations and businesses) construct the new structure of urban governance.

The citizens' patterns of behaviour both individually and collectively in the markets, civil society, and businesses of an urban region could be and are influenced by regulations such as: tax incentives, property rights. Also, direct interventions of local and national government in regards to welfare state benefits have an impact on it. In the meantime, market opportunities influence policies of the local government through encouraging divergence among tax incentives, regulations, and delivery of public goods (Sellers 2002). Therefore, the main objective of urban governance is to coordinate among actors with different values and consequently interests to solve collective action problems. May urban systems including but not limited to financial, political, public policy as well as information technologies, play crucial roles in functions of urban governance (Zwahr and Finger 2004).

The main challenge for urban governance is achieving the conflicting goal of enhancing accessibility to resources, security, and empowerment of citizens at the same time. While it seems like an impossible task for urban governing structure to resolve all social and political problems existing on and imposing by national level governance, it can certainly influence the circumstances and help to compensate for the insufficiency of national policies and assist its real stakeholders. A great opportunity lies beneath such challenges for effective urban governance to implement innovative solutions (Kearns and Paddison 2000). Creating a set of mechanisms that influences the dynamics among city actors and across different scales of economic, political and social aspects towards more partnership and collaboration could be a great help to the situation. In this context urban governance refers to the set of mechanisms for distributing power and allocating resources through the exercising control and coordination processes in a way that fosters partnership among all sectors of the society (Jessop 2002).

In fact, fairness in the form of creating capacities for all stakeholders of the society to grow and flourish is the essence of urban governance. Also, as stated eloquently, "good governance, seen as an integrated effort on the part of local government, civil society and the private sector" (Jessop 2002). To take advantage of the capacities of local participants, governing bodies require to create a local agenda according to which all local authorities and their communities are called upon to take part in a practical and sustainable plan (Huang et al. 1998; Varol et al. 2011). In order to keep such urban governance running smoothly, the system should provide all stakeholders the access to low-cost, understandable, and relevant information. The regulating restrictions, policies and laws should be also derived from analysing the same data sources. This way, the effectiveness of other governance indexes such as accountability and transparency is assured.

Urban governance should ideally be in continuous search for new mechanisms and methods that will effectively cultivate the resources of local communities and distribute them in a fair and effective way among the stakeholders. This is considered to be an effective way for governing urban systems as local communities have the capacity to explore and employ the relevant knowledge to local areas. Moreover, they can mobilize local institutional capacity, and cash out the social capital in order to resolve problems of the community and meet its needs. Social inclusion in these cases should be achieved with minimal state involvement or intervention. Accordingly, innovation as well as entrepreneurial economic activities will be supported and developed locally within such city governing structure. Finally, effective administration along with transparency in policymaking and execution of governmental responsibilities increases accountability when achieved through use of the local democratic process to make decisions responsive to the needs of residents (Kearns and Paddison 2000).

Urban governance for cognitive cities is very similar to the traditional models of it. The ultimate purpose should be serving all sub-systems (the same as stakeholders or agents or actors) and directly be affected by their outputs through feedback loops. In the case of a cognitive city: new methods need to be designed and new systems should be integrated to current one; all based on the latest technology that is well understood to meet the requirements and effectively installed within other urban infrastructures. These lead stakeholders to use the technology, relay on the system, and ideally participate in the governing system using smart devices including but not limited to computers. Governments need to encourage stakeholders to be part of cognitive city and contribute to the system through sharing data and participate in policymaking processes.

The significant point is to secure the shared data (particularly those who were shared with citizens) and to verify correct information is used in analytics and thus, making policies. The procedures for verification of authenticity, sanitation of data, and security of the anonymous information and ultimately knowledge bases, should also become a part of urban governance in cognitive cities. Only the secure data should be shared through city dashboards in public or through smart phone applications. When the body of urban governance could create, collect, and supply instant information flow for city actors and stakeholders, it has been elevated to intelligent or innovative governance system for a cognitive city.

Integrating intelligent urban governance for cognitive cities requires certain capacities. The city should first develop a budget plan, which will be partially invested in installation of city sensor as well as in securing channels of data collection through crowd sourcing. Some parts of the previous data collected by the city can be sanitized and used in primary stages of making policies. The installed city sensors will then be used to collect data on variety of behavioural matters such as patterns and flows of transportation in different modes, or patterns of energy consumption both in the house or business locations. Some visual sensors could also be used in helping citizens find parking spots in the city. Some others are used to monitor certain health indexes such as pollution level in the air, noise level, etc. Data collected from these sensors then get consolidated with data that is collected from citizens either through smart meters that will be an integrated part of all smart devices in the future, or from what they choose to contribute using the Internet via their smart phones and computers. The anonymous and sanitized data, when aggregated and merged in with sensor-based data would become a very reliable source of making smart and effective policies in cities of the future. In practice, ICTs provide an opportunity to create new social change and political structures. This change offers new horizontal communication networks that in turn define the characteristics of a network society. These network communities, which include electronically connected individuals and organisations, are smart communities' of the future (Khansari et al. 2016).

4 Current Concerns of Urban Systems

The world population is increasing and most this growth is happening in developing countries. Consequently, the environmental impacts of such growth are increasingly becoming an urban threat as well. Not only the population growth, but also the continued trends of overconsumption, hence, over pollution is considered a significant risk for the environment. Therefore, improving sustainabilty approaches for problems solving in urban level in on the rise. So are the efforts for finding ways to influence patterns of behaviour in the society. The increasing use of energy intensive products naturally gives rise to greenhouse gas emissions and accelerates climate change (Khansari et al. 2014a). The importance of the matter became apparent when many countries signed the two global protocols of Rio and Kyoto in 1992 and 1997 respectively to reduce the significantly negative impacts of direct emissions of CO_2 on climate change (Brechin 2003; Mathews 2007; McCright and Dunlap 2003).

Increasing the cost of environmental protection has changed the governmental environmental policy towards decentralized, flexible, self-regulated, and local environmental policies that encourage the participation of private sector in environmental regulation and reduce the role of the government through public access to environmental databanks (Khanna et al. 1998). The experience of sustainable companies confirms the profitability of these businesses due to the decrease of energy consumption and waste (Azapagic 2000; Jayne 2002). Individual behaviours leading to overpopulation and overconsumption constitute significant threats to the environment, and may also be magnifying the processes of global warming and ozone layer destruction. Therefore, behaviour change is required to move towards attainability. However, individuals do not always welcome behaviour change. The practice has shown that individuals in fact resist against such change (Khansari et al. 2014c).

Cognitive cities are capable of altering the environmental and social behaviours of citizens, whether this means providing information about mechanisms for reducing energy consumption, or updates on travel routes. Cognitive cities facilitate smart governance and political participation among citizens and officials through the use of ICTs like e-governance and e-democracy. They impact urban infrastructures such as systems of water and land use, energy, and transportation, and encouraging the use of renewable energy sources as a path to sustainable development. As shown in Fig. 1, a cognitive city can be divided into three layers, as follows:

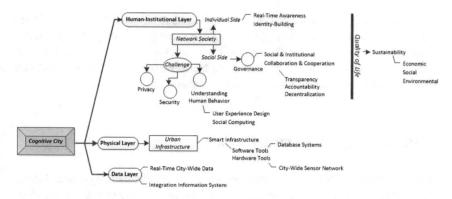

Fig. 1 The structure of cognitive city

- Human-Institutional Layer: Information technologies provide an opportunity to create new individual and social characteristics. This change offers new horizontal communication networks that in turn define the characteristics of a network society. In practice, network society that impacts on awareness and identity of individuals and governance improves social and institutional collaboration and cooperation. This results in more transparency, accountability, and decentralisation. Accordingly, the quality of life improves in network society with more economic, social, and environmental sustainability. On other hand, privacy, security, and understanding of human behaviour are main challenges of network society and user experience design and social computing are the tools that can be considered to deal with these challenges.
- Physical Layer: Smart infrastructure, a part of urban infrastructure, including software tools (e.g., database systems) and hardware tools (e.g., wired/wireless sensor network systems) shape network information technology systems in the physical layer.
- Data Layer: Real-time city-wide data and integration information systems form the data layer (Khansari et al. 2015).

Moving towards sustainability, among many other things, requires fundamental change of behaviour in regards to energy consumption and overall choices. To account for the social factor of collective behavioural patterns of citizens, here we only focus on individuals' energy behaviour in three levels: human-institutional, physical, and data levels. We investigate the role of information of smart systems on the energy behavioural change and present two integrated conceptual frameworks to increase the effective capacity of energy systems, facilitate energy consumption problem solving, and improve personal, social, and technical capabilities regarding energy consumption.

To address this aforementioned concern, our effort is to design a structure based on which the individuals' energy behaviour could be modelled. This will help us to investigate the parameters impacting the individuals' energy behaviour and the factors result in behavioural change. In order to do so, first we present a theoretical framework through which we discuss the importance of citizens' capability and their role in urban social networks in regards to changing energy consumption behaviours. In the following, we present the architectural structure of energy behaviour and consequently behavioural change discussing the three sub-layers that cumulatively cover the aspects of the households' energy consumption.

5 Influencing Behavioural Change

Human behaviours are systemic outcome of a multivariate function of: social structures, institutional contexts as well as cultural norms, among others. Assumingly, therefore, socio-structural networks and personal agency can highly impact the process of individual adaptation and change of behaviour. In practice, social structures form the rules and resources to organize, guide, and regulate actions of human agents. Meanwhile, human activities create, implement, and alter social systems (Khansari et al. 2014c). For instance, household environmental attitudes play an essential role in energy consumption behaviour. In other words, behaviour is affected by individual attitudinal variables and contextual variables such as interpersonal influences, regulations, interventions, institutional factors, incentives, constraints, knowledge and skills. In practice, moral norms, information, and communication significantly impact the energy consumption behaviours (Sapci and Considine 2014).

The relationships between individual contexts and attitudes, values, norms, and intentions affect the resilience of behaviours (Hobson 2003; Opschoor and van der Straaten 1993). In the process of energy saving, moral utility that identifies believes, values, attitudes, routines, norms, self-efficacy, constraints, and habits play a key role in social pressure for increased environmental sustainability. On the other hand, physical and structural conditions including home size, room size, widow area, technology, and standards positively influence information on energy consumption behaviour and residential households' demands for energy. A socio-demographic characteristic including households' income, age-compositions, and education level affects the residential energy consumption (Khansari et al. 2014a).

Financial incentives also affect individuals' energy consumption. In practice rewarding would encourage consumers to positively change their energy behaviours. Economics utilize rational choice theory to analyse behaviours of individuals. This theory points to the importance of the expected outcome of rational deliberation in forming consumption behaviours. According to this theory, self-concern is a basis of individuals' decisions. On the other hand, sociologists focus on the relationships between energy consumption and socio-technical systems. In practice, there is a bidirectional relation between social context and individuals' choice, that is, each affects the other one.

Social context plays a key role in defining the individuals' needs, attitudes, and expectations about social norms, technologies, infrastructures and institutions. Sociologists believe that the real energy consumption and the provision of energy resources do not determine individuals' decisions. In practice, "the difficulty for individuals to change their consumption patterns is also highlighted since lifestyle and use of material goods construct meanings and identities, which account for individual social expectations (social norms), self-expectation (positive or negative outcome of saving energy) and self-efficacy (perceived effort's effect to save energy)" (Pothitou et al. 2016). Both social norms and self-efficacy positively impact the reduction of energy consumption. Indeed, socio-technical forces affect individuals' habits, which in turn influence energy consumptions practices. Moreover, limitation of awareness about economic incentives (e.g., subsidies), limitation of capital incentives for energy-efficient equipment and limitation of knowledge about regulatory policies impact the individuals' energy behaviour change (Pothitou et al. 2016).

Technology is one of the other important factors in the process of consumption behavioural change and more precisely the decrease of energy consumption. For instance, smart meters can be used as an energy-monitoring tool to increase consumers' engagement and information on energy consumption. In practice, by utilizing energy meters householders have more control on their energy consumptions and costs (Khansari et al. 2014b). For example, capital-intensive petrochemical companies changes the fossil-fuel-based energy system to non-hydrocarbon energy technologies since fossil fuels are depleted and the costs of extracting fossil fuels are becoming too high. In practice, public authorities are able to transit smoothly to new environmental technologies to achieve sustainable energy in future (Kemp 1994).

6   The Proposed Energy Behaviour Structure

Population growth and continued trends of over-consumption and over-pollution represent significant threats to the environment, which necessitate a move towards more sustainable approaches. Moving towards sustainability requires individuals' energy behaviour change. Accordingly, this research focuses on individuals' energy behaviour in three layers: human-institutional, physical, and data layer.

Physical layer consists of all physical objects and infrastructures and their physical properties, and provides connectivity among the subsystems. For instance, wireless sensors can be installed in physical layer components to collect monitored parameters and transfer the data to the data network layer. Agents on the data network then use those sensors/actuators within the physical system to monitor the performance of city systems, and initiate control actions based on economic optimization considerations utilized within the social network layer. The human-institutional layer includes all individuals and organisations participating in the urban governance process. Stakeholders within this layer engage in both direct and indirect communication to develop policies, regulations, laws, and rules. The needs of all stakeholders are considered in the human-institutional level and stakeholders connect to the information systems of the data layer (Khansari et al. 2016). We investigate the role of information of smart systems on the energy behavioural change and present two integrated conceptual frameworks to increase the effective capacity of energy systems, facilitate energy consumption problem solving, and improve personal, social, and technical capabilities regarding energy consumption. Our developed structure for depicting the three layers of energy behaviours from agents' perspective is presented in Fig. 1. As shown in this figure, the capability of individuals and socio-structural network of interactions among them are the two main components of the human- institutional layer. Both of these components affect the individuals' decision-making process, as it is defined in this framework. However, these decisions may not be purely rational due to the limitations imposed by the bounded rationality, which is a result of unidentified complexities, uncertainty and their imposing risks (bounded rationality and rationality theories).

Economic, physical, and social processes shape consumption impacted by nature, environment, culture, laws, politics, and infrastructure of the society. The experience of the behaviour shape attitude and the behaviour is shaped by this attitude. Economics, demographics, consumer values, and psychology help to understanding of all aspect of consumption behaviour. Sociological perspectives complete marketings emphasis on economics and psychological perspectives consider consumption as a social process formed by cultural conventions and shared meanings, routines, cultural representations, and the tacit rules. This social structure governs behaviour in various social contexts.

Sharing information consider as a strategy that plays role in the behavioural change process. Educators, developmental psychologists, and social psychologists explain the process of knowledge translates to behaviour (Khansari et al. 2015). Energy consumption feedback strategy plays a key role in energy use reduction by 10–15% on average. Both energy consumption feedback and price strategies impact on reduction in energy consumption. Social norms are able to influence energy conservation through interactions between friends, neighbours, family in the community that emphasizes on energy saving. Energy cost, environmental attitudes, and social interactions affect energy conservation. To change energy behaviour, belief, value, and attitude should be considered. Combination of information and goal setting strategies is able to motivate individuals to have an energy efficient pattern. According to a psychological tendency, individuals look for supportive information to decrease their energy consumption. To achieve such goal, individuals need to improve their attitude and obtain required information (Khansari et al. 2014a).

The socio-structural network, however, affects the type of consuming energy, which could presumably be renewable or non-renewable energy, as presented in Fig. 2. In practice, lack of individuals' knowledge and capital are the main challenges that results in moving towards non-renewable energies. In fact, the cost of switching from fossil fuels to renewable energy sources is extremely high. Accordingly, the opposition argues that it is not reasonable to tolerate such high cost while the threats of climate change are still uncertain (Christiansen 2003). However, non-renewable energies are admittedly limited resources and using them also increases pollution, beyond any doubts. Indeed, to deal with this limitation and to decrease the pollution, either the energy consumption should be decreased or the efficient energy usage models should be followed.

Smart meter as a part of physical layer provides a bidirectional communication between energy consumer and energy provider, encourages customers to follow effi-

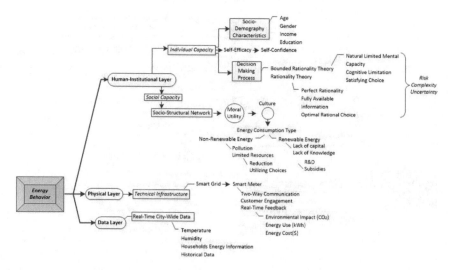

Fig. 2 The structure of energy behaviour

cient energy usage models, and provides real-time information about energy usage, energy cost, and environmental impact of CO_2. Provision of this information encourages the consumers to reduce their energy consumption. Figure 2 depicts that data layer provides real-time citywide data including: temperature, humidity, the average of households' energy information, and historical average energy usage of the city. This information in turn decreases the energy consumption.

7   A Governing Framework for Change

To encourage the individuals to follow efficient usage model, their energy behaviour should change. Figure 3 shows our developed architecture depicting the process of behavioural change. As shown in this figure, this architecture includes human-institutional, physical, and data layers. This chapter focuses on the Human-institutional layer. As depicted in Fig. 3, human-institutional layer shows that individuals and their social networks can highly influence energy behavioural change. In practice, there is a resistance against this change rooted in the individuals' culture and their capabilities. To deal with this resistance, a number of strategies can be applied including price, goal setting, socio-structural, comparative feedback, and information strategies.

Economic, physical, and social processes shape consumption impacted by nature, environment, culture, laws, politics, and infrastructure of the society. The experience of the behaviour shape attitude and the behaviour is shaped by this attitude. Economics, demographics, consumer values, and psychology help to understanding of all aspect of consumption behaviour. Sociological perspectives complete marketings

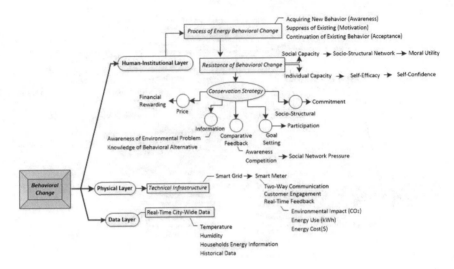

Fig. 3 The process of behavioral change

emphasis on economics and psychological perspectives consider consumption as a social process formed by cultural conventions and shared meanings, routines, cultural representations, and the tacit rules. This social structure governs behaviour in various social contexts. In the process of energy saving, moral utility that identifies believes, values, attitudes, routines, norms, self-efficacy, constraints, and habits play a key role in social pressure for increased environmental sustainability. On the other hand, physical-structural conditions including house size, room size, widow area, technology, and standards positively influence information on energy consumption behaviour and residential households' demands for energy. Socio-demographic characteristics including households' income, age-compositions, and education level affect the residential energy consumption (Khansari et al. 2015).

These Sects. 6 and 7 present two comprehensive frameworks to address the households' energy consumption behaviour and energy behavioural change. These frameworks provide a broad overview of key agents affecting energy consumption including personal, social, and technical aspects. In other words, we consider individual and social structure of the energy behaviour and show the role of information of smart energy systems on the process of energy behavioural change.

CO_2 emissions are considered to be the main basis of the incredible increase in the earth's surface temperature in recent years. Most emissions result from human activities. Thus, developing a detailed framework representing the parameters affecting individuals' energy behaviours is required. This chapter offers an integrated conceptual framework to increase the efficiency of energy systems under complex and uncertainty conditions, facilitate energy consumption problem solving, and support the development of capacities at the individual, social, and technical levels to improve managing energy consumptions in the future. This research presents a conceptual soft systems model to explore the process of individuals' energy behaviour change based on socio-structural and techno-structural contexts. The proposed model provides a broad overview of the key agents affecting energy consumption. The model is created based on the research in the literature discussing the causal relations between various variables. In practice, this chapter maps the impact of the information technologies on the decision-making process. The authors emphasise the essential role of information technologies in improving decision-making through decentralisation and connectivity, and leading communities to become smarter by providing data to stakeholders, including officials and agencies in cognitive city.

Involving the implementation of sustainable technologies and the deployment of information availability play essential roles in reducing environmental impact by positively affecting individual energy behaviour. The implementation of electronically-based governance shifts political participation to the web and cellular networks and enables superior feedback and community involvement. This socio-technical systems modelling is necessary to cope with the continuous and prolific stream of data collection that has reshaped how cities and structures of governance are designed. The models have also provided how information technologies offer ways for residents to play a role in local governance, facilitate sustainable urban infrastructure, and reduce household energy consumption. Developed models describe how practices of information sharing and different structures of governance and management impact the willingness of individuals and households to adopt sustainable energy practices.

8 Conclusion

This chapter is dedicated to providing the readers with a governing framework that can impact citizens collective behaviour in regards to energy consumption by flow of information and learning processes. This could be of interest for academicians as well as city policymakers, executives, entrepreneurs, and enthusiastic readers who are seeking embedded cognitive capacities as a way of implementing sustainability and resilience in urban systems of future. The basic assumption in this research is that stream of timely information could influence decision-making and behavioural patterns in long-term.

We are in an increasing slope of urbanization in history. This has been a trend for decades and as a result more than half of our planet is consist of cities, which create more than half of global GDP. Such economic impact includes many other critical factors such as consumption of resources and creation of pollution. These challenging circumstances however, create opportunities for addressing issues effectively at the same time. That is why urban systems play a pivotal role in achieving sustainability, resilience, and efficiency. In better words, the same systemic forces that created many of our global issues might be a start point for an evolutionary process that not only survives the challenge but also helps us excel in our collective achievements. This chapter indicates on ways in which effective governance can impact on sustainability through cognitive processes provided by information and communication technologies. An informed decision-making process, when streamed in real-time can increase policy effectiveness, accountability, and transparency within urban systems and enhance citizens behavioural pattern towards sustainability values. Based on this premise, a conceptual system for the cognitive city and energy behaviour including three sub-layers as human-institutional, physical, and data layers was elaborated and the impact of such integrated conceptual models to increase the efficiency of energy systems was discussed. The systems methodology, which was adopted in this chapter, shows how sharing the right information and an effective structure of governance can incentivize citizens to adopt sustainable energy practices.

To achieve this purpose, cities are in need for more cognitive computing systems. Such systems can facilitate effective and real-time communications among citizens and more importantly, creates a platform for policymakers to adjust their approach in a timely manner. It also requires more effective learning processes through analysis of collected data and cognitive computing systems. An effective education system for citizens to bring them up to speed with learning capabilities of the city should be considered and facilitated as well. Given a responsive sensing system, a powerful computational system at the city level, and informed citizens who are familiar with technology to an extent that equips them with means for interaction, we can achieve the ultimate goal of cognitive cities that is making the right decisions both by governing bodies and citizens towards a sustainable and resilient urban environment.

References

Azapagic A, Perdan S (2000) Indicators of sustainable development for industry: a general framework. Process Saf Environ Protect 78(4):243–261

Bache I, Flinders M (2004) Multi-level governance and the study of the British state. Public Policy Admin 19(1):31–51

Batty M (2013) The new science of cities. MIT Press, Cambridge

Bevir M (2010) Democratic governance. Princeton University Press, Princeton

Brechin SR (2003) Comparative public opinion and knowledge on global climatic change and the kyoto protocol: the US versus the world? Int J Sociol Soc Policy 23(10):106–134

Chaskin RJ, Abunimah A (1999) A view from the city: local government perspectives on neighborhood-based governance in community-building initiatives. J Urban Affairs 21(1):57–78

Christiansen AC (2003) Convergence or divergence? Status and prospects for US climate strategy. Clim Policy 3:343–358

Darabi HR, Gorod A, Mansouri M (2012) Governance mechanism pillars for systems of systems. In: 2012 7th international conference on system of systems engineering (SoSE). IEEE, pp 374–379

Darabi HR, Mansouri M, Gorod A (2013) Governance of enterprise transformation: case study of the FAA NextGen project. In: 2013 8th international conference on system of systems engineering (SoSE). IEEE, pp 261–266

Finger M, Pécoud G (2003) From e-government to e-governance? Towards a model of e-governance. In: Proceedings of the 3rd European conference on e-government-ECEG, MIR-ARTICLE-2005-061, pp 119–130

Green GP, Haines A (2011) Asset building & community development. Sage Publications Inc, Thousand Oaks

Healey P (2004) Creativity and urban governance. Policy Stud 25(2):87–102

Hobson K (2003) Thinking habits into action: the role of knowledge and process in questioning household consumption practices. Local Environ 8(1):95–112

Huang SL, Wong JH, Chen TC (1998) A framework of indicator system for measuring Taipei's urban sustainability. Landsc Urban Plan 42(1):15–27

Jacobs J (1961) The death and life of great American cities. Random House, New York

Jacobs J (1969) The economy of cities. Random House, New York

Jayne V (2002) Triple bottom line reporting-being seen to give a damn. N Z Manag 49(6):60–61

Jessop B (1998) The rise of governance and the risks of failure: the case of economic development. Int Soc Sci J 50(155):29–45

Jessop B (2002) Liberalism, neoliberalism, and urban governance: a state-theoretical perspective. Antipode 34(3):452–472

Kearns A, Paddison R (2000) New challenges for urban governance. Urban Stud 37(5–6):845–850

Kemp R (1994) Technology and the transition to environmental sustainability: the problem of technological regime shifts. Futures 26(10):1023–1046

Khanna M, Quimio WRH, Bojilova D (1998) Toxics release information: a policy tool for environmental protection. J Environ Econ Manag 36(3):243–266

Khansari N, Mostashari A, Mansouri M (2014a) Conceptual modeling of the impact of smart cities on household energy consumption. Proc Comput Sci 28:81–86

Khansari N, Mostashari A, Mansouri M (2014b) Impact of information sharing on energy behavior: a system dynamics approach. In: 2014 IEEE international systems conference (SysCon)

Khansari N, Mostashari A, Mansouri M (2014c) Impacting sustainable behavior and planning in smart city. Int J Sustain Land Use Urban Plan (IJSLUP) 1(2):46–61

Khansari N, Darabi HR, Mansouri M, Mostashari A (2015) Case study of energy behavior: systems thinking approach. In: 2015 IEEE international systems conference (SysCon)

Khansari N, Finger M, Mostashari A, Mansouri M (2016) Conceptual systemigram model: impact of electronic governance on sustainable development. Int J Syst Syst Eng 7(4):258–276

Lodge M (2010) Key concepts in governance-by Mark Bevir. Public Admin 88(4):1143–1145

Lummis CD (1996) Radical democracy. Cornell University Press, Ithaca

Mansouri M, Mostashari A (2010) A systemic approach to governance in extended enterprise systems. In: 2010 4th Annual IEEE systems conference. IEEE, pp 311–316

Mathews J (2007) Seven steps to curb global warming. Energy Policy 35:4247–4259

McCright AM, Dunlap RE (2003) Defeating kyoto: the conservative movement's impact on US climate change policy. Soc Probl 50(3):348–373

Mostashari A, Arnold F, Mansouri M, Finger M (2011) Cognitive cities and intelligent urban governance. Netw Ind Q 13(3):4–7

Opschoor H, van der Straaten J (1993) Sustainable development: an institutional approach. Ecol Econ 7(3):203–222

Osborne SP (2010) The new public governance: emerging perspectives on the theory and practice of public governance. Routledge, London

Pierre J (1999) Models of urban governance: the institutional dimension of urban politics. Urban Aff Rev 34(3):372–396

Pierre J (2005) Comparative urban governance uncovering complex causalities. Urban Aff Rev 40(4):446–462

Pothitou M, Kolios AJ, Varga L, Gu S (2016) A framework for targeting household energy savings through habitual behavioural change. Int J Sustain Energy 35(7):686–700

Rhodes RA (1997) Understanding governance: policy networks, governance, reflexivity and accountability. Open University Press, Buckingham

Rossel P, Finger M (2007) Conceptualizing e-governance. In: Proceedings of the 1st international conference on theory and practice of electronic governance. ACM, pp 399–407

Sapci O, Considine T (2014) The link between environmental attitudes and energy consumption behavior. J Behav Exp Econ 52:29–34

Sellers JM (2002) The nation-state and urban governance: toward multilevel analysis. Urban Aff Rev 37(5):611–641

Shahin J, Finger M (2008) The operationalisation of e-governance. In: Proceedings of the 2nd international conference on Theory and practice of electronic governance. ACM, pp 24–30

Varol C, Ercoskun OY, Gurer N (2011) Local participatory mechanisms and collective actions for sustainable urban development in Turkey. Habitat Int 35(1):9–16

Zwahr T, Finger M (2004) Critical steps towards e-governance: a case study analysis. In: European conference on electronic government, MIR-CONF-2005-016

Zwahr T, Rossel P, Finger M (2005) Towards electronic governance: a case study of ICT in local government governance. In: Proceedings of the 2005 national conference on Digital government research, digital government society of North America, pp 53–62

Mo Mansouri received the B.Sc. degree in Industrial Engineering from the Sharif University of Technology, Tehran, Iran, the M.Sc. degree in Industrial Engineering from the University of Tehran, Tehran, and the Ph.D. degree in Engineering Management from The George Washington University, Washington, DC, USA. He is Research Associate Professor and Program Lead in Systems Engineering & Socio-Technical Systems with the School of Systems and Enterprises, Stevens Institute of Technology, Hoboken, NJ, USA. Prior to joining Stevens, he served at several international development and nonprofit organizations as a Research Fellow or Consultant, working on strategic philanthropy and social entrepreneurship for development programs. He publishes in many scientific journals and on a range of domains from transportation to financial and energy systems (among others: the IEEE Systems Journal, the International Journal of Industrial and Systems Engineering, the Journal of Operational Risk, Marine Policy, Maritime Policy and Management, the International Journal of Ocean Systems Management, Enterprise Information Systems, Transportation Research Record) and in numerous conferences, particularly the IEEE Systems Conference. His current research focuses on developing governance frameworks for effective policymaking and embedding resilience in complex networks and infrastructure systems.

Nasrin Khansari received her B.A. degree in Economics from Allameh Tabataba'i University, Tehran, Iran as well as an M.A. degree in Social Science from New York University, New York, NY, USA, and in Social Development from the University of Tehran, Tehran, Iran. She holds a Ph.D. degree in Systems Engineering from the Stevens Institute of Technology, Hoboken, NJ, USA. Currently, she is a Postdoctoral Researcher within the Department of Electrical & Systems Engineering at the University of Pennsylvania, USA. In this position, she conducts research related to complex adaptive systems and aid in the design of agent-based decision support systems with most of her work focused on transportation energy consumption behaviors. Her research is on the role of information technologies, as a way for residents to contribute to the decision-making process and focused on the mechanisms by which information sharing enables overcoming bounded rationality problems within urban governance. She also works on an integrated conceptual framework to address the issue of CO_2 emissions, increase the efficiency of energy systems, facilitate energy consumption problem solving, and support the development of capacities based on socio-structural and techno-structural contexts.

Extending Knowledge Graphs with Subjective Influence Networks for Personalized Fashion

Kurt Bollacker, Natalia Díaz-Rodríguez and Xian Li

Abstract This chapter shows Stitch Fix's industry case as an applied *fashion* application in cognitive cities. Fashion goes hand in hand with the economic development of better methods in smart and cognitive cities, leisure activities and consumption. However, extracting knowledge and actionable insights from fashion data still presents challenges due to the intrinsic subjectivity needed to effectively model the domain. Fashion ontologies help address this, but most existing such ontologies are "clothing" ontologies, which consider only the physical attributes of garments or people and often model subjective judgements only as opaque categorizations of entities. We address this by proposing a supplementary ontological approach in the fashion domain based on subjective influence networks. We enumerate a set of use cases this approach is intended to address and discuss possible classes of prediction questions and machine learning experiments that could be executed to validate or refute the model. We also present a case study on business models and monetization strategies for digital fashion, a domain that is fast-changing and gaining the battle in the digital domain.

Keywords Ontology · Folksonomy · Knowledge graph · Fashion · Subjectivity
Temporal networks · Social networks · Influence · Natural language processing
Recommendation systems · Personalization · Business development

K. Bollacker · N. Díaz-Rodríguez (✉) · X. Li
Stitch Fix Inc, San Francisco, CA, USA
e-mail: ndiaz@decsai.ugr.es
URL: http://multithreaded.stitchfix.com/, http://www.stitchfix.com

K. Bollacker
e-mail: kbollacker@stitchfix.com

X. Li
e-mail: xli@stitchfix.com

E. Portmann et al. (eds.), *Designing Cognitive Cities*, Studies in Systems,
Decision and Control 176, https://doi.org/10.1007/978-3-030-00317-3_9

1 Introduction

1.1 Cognitive Cities and the Fashion Domain

Cognitive computing aims at improving the quality of life in cities, especially aiding in decision-making, handling linguistic information—which is usually imprecise and developing applications towards achieving Smart Cities (D'Onofrio and Portmann 2017).

This chapter shows a use case beyond the cognitive cities, through a case on cognitive-cultural capitalism emanating from the new urbanism (Scott 2014). Because creativity is a concept whose time has come in economic and urban geography, and because much existing research on creative cities fails adequately to recognize that interdependent processes of learning, creativity and innovation are situated within concrete fields of social relationships (Scott 2014), we show a use case where data science and cognitive computing are applied to a successful and real business case. In spite of such modern urban mechanisms, which *tend to offer a flawed representation of urban dynamics* and lead sometimes to *essentially regressive policy advocacies*, cognitive-cultural capitalism is a robust theoretical framework through which contemporary urbanization processes can be described, as it is well motivated in (Scott 2014). This framework, and our use case concretely, show to have larger impacts on urban outcomes leading to higher degree of automation and job transformations.

Because of the AI transformation we are living in, we demonstrate how it is important to combine analytics, NLP, Web knowledge, fuzzy (cognitive maps), etc. with the fashion industry to develop new experiences in smart/cognitive cities that adapt to the citizens' new lifestyle.

1.2 Stitch Fix's Case

Stitch fix (SF) is a personalization company that delivers fashion to your door and it is purely driven by data science (Fig. 1). Stitch Fix is listed as number 2 in the top 15 companies to watch in 2017,[1] and has been popular in the US, now running for 5 years, since CEO Katrina Lake tested the viability of the idea in her apartment in Massachusetts. The company counts with a peculiar title of Chief Algorithms Officer Eric Colson (and a subset of the team) that comes from the heavy recommendation system-based company Netflix. Stitch Fix is thus sometimes known as the "Netflix of fashion","Pandora for clothing", or the "clothing company that blends AI and

---

[1]Inc: Top 15 companies to watch in 2017 http://www.inc.com/guadalupe-gonzalez/ss/top-companies-to-watch-2017.html.

Fig. 1 Stitch Fix service

Human Expertise".[2] In the San Francisco Bay Area, Stitch Fix is popular as well for having a thorough curated blog on how they tackle different problems with data science (see Multithreaded series blog[3]).

Competitive asset: Why is Stitch different to the rest of online fashion retailers?
Big fashion retailers in US such as Macy's, Nordstrom and Gap continue to lay off employees and shutter stores in the face of increased competition from online players. Stitch Fix, is not characterized for having a fast delivery of fashion at home, nor having very popular nor high price point luxury brands; they have more exclusive vendors, as well as their own unique brands, to provide a more personalized experience to the client's profile, budget, requests, and lifestyle. As Colson says, "We're not going to be better priced, or faster shipping (their delivery is not supposed to be really fast), or a better brand, so we have to be really good at relevancy."

At Stitch Fix there is no online catalog, and the customer does not choose what is getting in a box. The aim is to have effortless shopping taken care of by a personal stylists that hand-picks a look for you. You can try in the comfort of home, keep what you like and return what you do not, for free, and giving feedback on how the stylist did for better improvement in next delivery. The personalization and styling experience is supported with social networks information such as the information

---

[2]Harvard Business Review https://hbr.org/2016/11/how-one-clothing-company-blends-ai-and-human-expertise.

[3]Multithreaded series blog http://multithreaded.stitchfix.com/.

Fig. 2 Stitch Fix

that users add to their Pinterest boards, in order to give the stylist a visual glimpse of liked garments.

Stitch Fix is not distinguished by high prices, its main products are within an affordable budget versus other services such as *Trunk Club*,[4] which targets larger budget customers. Stitch Fix focuses on partnership, and cross collaboration, "together we are better" is a motto, and customer experience is the priority: if a feature does not help the client, they do not go for it. Stylists are recognized, and can work flexibly part time from home. The objective is enhancing the quality of their products and get standard procedures in place (Fig. 2).

2 Beyond Clothing Ontologies: Modeling Fashion with Subjective Influence Networks

If Stitch Fix's personal stylists were to search for the perfect outfit for one of our customers who is asking advice for an outfit for his first Burning Man attendant or for a 50 year wedding anniversary party, it would be ideal if we could search in our database for styles such as rockabilly, retro, boho or "best pieces from the 70's". However, despite having ample of physical attribute annotations from different vendors, as well as algorithmically generated garments, we still do not have a sufficient purchase on the abstract attributes of a consumer's perception, such as the aforementioned categories.

We address this problem by proposing a supplementary ontological approach in the fashion domain based on subjective influence networks. They measure novelty,

---

[4]https://www.trunkclub.com.

impact and represent influence mechanisms that can validate or refute hypotheses that contain subjective or aesthetic components. The components of an influence vector instance consist of the elapsed time between the beginning of styles, the magnitude of the influence, the mechanism of influence, and the agent of influence.

We enumerate a set of use cases this approach is intended to address and discuss possible classes of prediction questions, hypothesis testing and machine learning experiments that could be executed to validate or refute the model. For instance, using network mapping, we could find answers to: *Can influences in other cultural domains such as music be used to predict fashion influences?* Other application is quantifying subjective attributes, for instance, subjective influence networks could characterise the differences between retro and classic glasses and their mechanisms of influence.

Example hypotheses to confirm/reject could be: *Are retro glasses those worn by 2 generations back? Are classic glasses those that never go out of fashion?* Other example of use case is predicting fashion cyclicality, and so, we could postulate: *Can influence networks tell apart fashion cycles periodicity to predict when bell trousers will be fashionable again?* Representing fashion evolution on social media (e.g. hashtags on *Instagram*) as influence mechanism is an example to measure influence within time and space scales to evaluate the "viral" nature of rapid style changes. Through the use of subjective influence networks, we plan to augment the Knowledge Graph of fashion information, a search engine that allows advanced faceted search to infer features even if they are not explicitly tagged in the merchandise. The final aim is understanding and translating into machine-consumable manners the way humans perceive and transmit aesthetics and style in fashion, and more generally, subjectivity, something that computers do not (yet) do better than us.

3 Background: Ontologies and Knowledge Graphs

As on-line fashion retail industry has been growing rapidly against traditional physical shopping, there has been a corresponding shift to a much more data-driven paradigm for business operations including manufacturing, merchandising, and marketing. In particular, future-focused data analysis has become a particularly important activity, such as predicting fashion trends, price forecasting, construction of recommender systems, and identification of consumer influencers. Often, these activities are approached using statistical, machine learning or other data-driven techniques. However, much of the data in the fashion domain comes from deep, diverse, cultural entities and phenomena. While fashion in itself is part of and can define culture, it also borrows from other cultural domains, such as music, language, film, religion, mythology, local folklore and many others. In most cultural domains, it is important to understand the narrative of history and contemporary subjective judgements and opinions. For example, in music, Italian words are used to contextualize abstract musical concepts (e.g., *allegro, largo, presto*). However the meaning of these words in the context of music has evolved and diverged from their original, common definitions. Knowing the history as well as the current interpretation of these words by the composers who use them is required to fully understand their musical meaning. Similarly, fashion is an inherently subjective, cultural notion. It is defined not by quantitative, testable measures, but by its history and the perceptions of people who care enough to form opinions about it. Therefore, in order to understand fashion in any rigorous way, this subjectivity must be an intrinsic part of the model.

One of the techniques for addressing the subjective, cultural parts of a knowledge domain is to use ontologies. Schemas, ontologies and its data population through knowledge graphs (KGs) are formal tools for expressing organized meaning and provide sense or context to a domain. More concretely, ontologies often integrate common-sense and human expert knowledge as well other external knowledge sources into machine readable computational models. Unfortunately however, most existing ontological work in fashion partially avoids subjectivity by simply focusing on "clothing ontologies" rather than fashion as a whole. Clothing ontologies primarily model the structure of physical feature values (e.g., sleeve length, colors, fabric). A particular garment can be represented in a multidimensional feature space chosen from such an ontology. Usually each garment class (e.g., top, bottom, shoe, hat) is considered to have a distinct feature space from other classes. When they do include subjective elements, clothing ontologies often do this through the inclusion of non-objective features (e.g., expected occasion, style category), but these features are usually opaque categorizations of entities, with no explicit semantics. Despite the limitations, these clothing features spaces are still useful because they provide semantic structure to data that can be used when applying analytic/prediction techniques (e.g., similarity measures, classifiers, function estimators).

We believe that deeper, richer representations of the subjective features of fashion data is possible and would help in many important use cases. In this paper, we propose an architectural augmentation to traditional clothing ontologies that includes the notion of a subjective influence network in a way that may be able to capture subjective semantics that simple categorical features do not. We enumerate a set of potential use cases, and propose types of measurements and applications that can be carried out to measure the usefulness of our approach.

The rest of the paper is organized as follows. Section 4 exhaustively summarizes the state of the art on existing fashion ontologies and frameworks and Sect. 4.4 describes machine learning applications as motivating use cases for our fashion ontological modelling approach. Section 5 proposes the theoretical foundations of the subjective model of influence, entities, relations and the mechanisms to quantify influence and subjectivity. Section 6 discusses evaluation approaches and utility of the model once populated with empirical data. Section 9 concludes with further insights.

4 Related Work

4.1 Related Fashion Ontologies and Schemas

Ontologies have been used to represent knowledge in a large set of real-life problems, from genetics[5] to decision support systems, optimization, matchmaking and human activity recognition (Díaz-Rodríguez et al. 2014c). In the fashion world, ontologies have sporadically been used for recommendation systems. For example, ontologies have been combined with fuzzy logic for personalized garment design, where fuzzy decision trees serve in learning a set of representative samples. Fuzzy cognitive maps model complex relations between sensory descriptors and fashion themes given by consumers to provide more fine grained recommendations as well as the evaluate how much a specific body shape is relevant to a desired emotional fashion theme (Wang et al. 2015).

An important existing ontology is the Garment Style Advice Ontology SERVIVE (SERVice Oriented Intelligent Value Adding nEtwork for Clothing-SMEs embarking in Mass- Customisation)[6] (Vogiatzis et al. 2012). The Servive Fashion Ontology (SFO) includes relations among different categories of entities such as *colors, companies, garment features, materials*, etc. and provides a similarly structured and unified vocabulary to represent human, fashion and manufacturing concepts. The project includes the design of a Virtual Customer Advisor (VCA) which expresses preferences for a given garment that is evaluated via SWRL rules and Pellet reasoner. Figure 3 shows the most abstract or top layer classes as well as the highest hierarchical layer of object properties modelled in SERVIVE ontology. Despite being the most complete ontology publicly available to the best of our knowledge, except for the subjective season labels (*hasHumanStyleColour*) and suitability classifications (*isForOccasion*), the ontology consists only of physical object hierarchies.

Ontologies per se act primarily as a modelling tool, and for them to be useful, they are to be integrated into some kind of application (be it search, recommendation, classification or decision making applications). For instance, ontologies have also been integrated into probabilistic and media-rich approaches for personalized garment recommendation systems. Expert subjective knowledge from public online media is used to compute compatibility among products and user profiles according to context and probabilistic reasoning. (Ajmani et al. 2013) concretely focuses on dresses (*sarees*) and its evaluation of several individuals' fashion preferences and celebrities' actual choices compared with automated recommendations. The format of the ontology is MOWL, that *enables the analysis of visual properties of garments with respect to fashion concepts*, but it is not publicly available.

Another ontology, which considers designers, models, trends, seasons and celebrities is in (Novalija and Leban 2013), which exploits lexico-syntactic patterns as NLP

---

[5]http://geneontology.org/.
[6]SERVIVE EU Project http://www.servive.eu/.

Fig. 3 SERVIVE Ontology
main entity classes (above)
and main object properties
(below) (Vogiatzis et al.
2012)

tools for ontology learning, relation extraction and curation through domain experts.
Table 1 summarizes the main ontologies' concepts and relations modelled.

Considering work that is more general than the fashion domain, open data portals
such as Dbpedia and Freebase (Bollacker et al. 2008) contain 1 K topics and 3 K facts
around fashion, clothing and textiles.[7] Despite the richness and structure found in

―――――――――――

[7] https://developers.google.com/freebase/.

Table 1 Existing clothing ontologies

Ontology/Model and language	Main entities	Relations
SERVIVE (Vogiatzis et al. 2012), OWL	Body type, colors, companies, garments (features, material), human colour categories, seasonal human style color, occasion, style	Co-occurs, hasInterest, hasHigh/Low/NeutralRecommendation. hasBodyType/Fit/EyeColour, hasGarmentButtons/Colour/Feature/Material/HumanStyleColour, hasOcassion, hasSleeves/Stripes/Style/styleDescription, isForOccasion, manufacturedBy, similarTo, isColour
Fashion ontology (Novalija and Leban 2013), RDF	Celebrity, designer, model, clothing term, trend, season	
Indian garment ontology (Ajmani et al. 2013), MOWL	Craft (stitch, print, embroidery), material, textile categories,	Celebrity validation
Fashion cognitive model (Li and Li 2012)	Garment parts (silhouette, waist, length, collar, sleeve, ornaments, symmetry)	
Fashion cognitive model (Wang et al. 2015)	Body shape, desired emotional theme	Effectiveness, acceptability, realizability

these formal base resources, the creative and subjective, contextual part of fashion is missing from these knowledge bases.

4.2 Cognitive Models for Fashion Modelling

In the literature there are non-ontological models which frame similar problems. They blend human and machine models for evaluating specific body shapes' relevance to a desired emotional fashion theme or intention to be transmitted. For instance, in (Wang et al. 2015), *effectiveness* evaluates whether recommended styles are relevant to the design objective or desired fashion theme, *acceptability* refers to whether the best recommended style is accepted by the expert, and *realizability* assesses if the proposed recommender system can be applied to the fashion (Wang et al. 2015).

An example of a cognitive model for fashion style decision making is in (Li and Li 2012), where Genetic Algorithms enhanced with Multi-alternative Decision Field Theory (MDFT) tackle the context and choice set problem in decision making by using *psychological distance* between alternatives. The latter is based on the Euclidean distance among positions in a multi-attribute-dimensional subjective evaluation space.

4.3   Subjectivity in Other Domains

We identify a lack of a *subjective style* schema in the related work that goes beyond the biology or mechanics of clothing, and that expresses a more wholistic personal approach than the existing *inventory* clothing ontologies. By *inventory ontologies*, we mean those based on static attribute-based or physical feature spaces.

Other subjective and hard to describe domains such as music also benefit from having taxonomical classifications in form of ontologies. For instance, projects such as MusicBrainz[8] collects music metadata, and the Music Ontology[9] (Raimond et al. 2007) is a formal framework to deal with music-related information on the Semantic Web including editorial, cultural and acoustic information. Just like in music, a fashion ontology can integrate fashion-related data across multiple sources, or enrich search-engine results around decades, styles or influencers. Because of this, musicians might be useful allies for the fashion industry, (e.g., *thanks to their status as bohemian individuals*) and music industry might need fashion (Miller 2011), e.g., to model music taste or predict fashion cliques.

Another similar natural phenomenon is language, where influence networks, among many other factors in time, model organically the evolution of its spread, its vocabulary, grammar rules, tonality, etc. In all, music, fashion and languages, influence and subjectivity are inherent to the domain and for them to fully be considered into machine learning systems, they need to be modelled quantitatively.

4.4   Fashion Ontology Use Cases

In fashion, the human component of algorithm evaluation is necessary (Steinfeld et al. 2007; Wah et al. 2011). Guided by this, we identify candidate applications where a fashion ontology enhanced with a better subjective data representation would likely be helpful.

1. Defining stylistic rule guides and recommendations or predicting specific trends. For instance, to answer questions on: how to be edgy and ahead of the fashion trend without being too far off, or how to predict the Oscars' ceremony outfits?.[10]
2. Predicting mass production trends. For example, the problems of cost-efficient budget and resource allocation as well as market demand optimization.
3. Providing organizing structure, e.g., taxonomy or folksonomy, for fashion annotation systems that leverage crowd-sourced online data (e.g., Zoghbi et al. 2013, 2016).

---

[8]The Open Music Encyclopedia https://musicbrainz.org/.

[9]http://www.musicontology.com.

[10]http://www.usatoday.com/story/life/entertainthis/2016/02/23/oscar-fashion-predicting-what-stars-wear-red-carpet/80747356.

Fig. 4 Traditional clothing
feature space

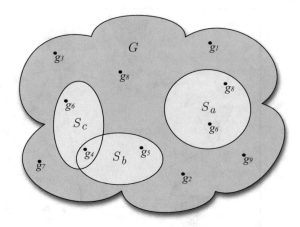

In next section we specify our augmentation for clothing ontologies, including a description of modeling obligations needed to make it useful and examples of first order measurements of represented data.

5 Modeling Subjective Influence

5.1 Styles as Regions in a Feature Space

So how can subjectivity semantics be modeled as an influence network? Let us first consider a somewhat traditional interpretation of features of garments, based on physical properties from a clothing ontology. A particular garment g can be represented as a point in a clothing feature space G (see Fig. 4). Let there be a theoretical set of all clothing styles Φ such that $\forall g \in G$ a subjective judge function $s()$ assigns a classification $s(g)$ such that $s(g) \in \Phi$. We define a distinct "style" x to be a region $S_x \subset G$ such that $\forall g \in S, s(g) = x$.

Because the S_x depends only on a single subjective function $s()$, it does not consider the fact that for any x, there may be multiple subjective functions that are contradictory. However, we believe this reflects the actual messiness of the real world.

5.2 Styles in a Network

Styles represented as a collection of points in a physical clothing feature space do little to capture the (subjective) semantics of fashion beyond the opaque categorical $s(g)$ features. To capture richer semantics, first consider a style as the human perception of the physical features of a single garment (or entire style region). Each

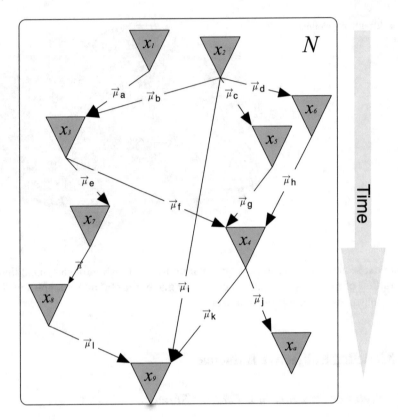

Fig. 5 Influence Network

style can then be described as a coherent aesthetic entity in the mind of an observer. While traditionally this style may be quantitatively described by its physical features, consider the alternative aspect of its subjective qualities shared with other cultural entities. These entities could be other clothing styles, or could be from other cultural domains external to fashion (e.g., music, sports, film, art, literature). We model this subjectivity as a network of influence.

We treat each style x as a node in an acyclic graph/network N (see Fig. 5) such that there is a temporally directional edge function $e(x, y)$ that specifies the influence between nodes. Moving backward in time (y to x), an edge between styles describes the stylistic borrowing that occurs. Moving forward in time (x to y), the edge represents the influence from older to newer styles. This influence is not a single measure, but rather a collection of influences of different mechanisms. The strength of each mechanism can be represented as a single positive number. More formally:

$$\forall x, y \in \Phi, \exists e(x, y) \tag{1}$$

such that $e(x, y) = \overrightarrow{\mu_{xy}}$ where $\overrightarrow{\mu_{xy}}$ is the influence vector from x to y,

Each element of $\vec{\mu}$ can be treated as a quad (t, i, m, a) where t is the amount of elapsed time between the influencing and influenced style, i is the intensity or strength of influence, m is the mechanism of influence, and a is the agent of influence. While t and i can both be represented as positive reals, m is a class that exists in a (likely) vast space of possible mechanisms M. Some categories of $m \in M$ might be:

- *Explicit*: The creator of a style explicitly declares previous styles that have been influential in the current creative process.
- *Calculated*: Algorithmic or other mechanical means may estimate influence mechanism and strength based on garment features or causal cultural models.
- *Extrinsic*: The influence may be caused by cultural influences in one or more external parallel cultural influence networks (e.g music, religion, sports) For example, a musician that borrows musical style from a revered earlier musician, may also borrow fashion elements for their own public image.

There are many different types of possible agents of influence a, including:

- *Well known individual persons or small groups*: These extrinsic influencers may be fashion designers, well known artists/performers, cultural icons, or celebrities who are admired for the artistic or political talents.
- *Organizations*: Corporations whose business is in the fashion create styles and attempt to maximize the desirability of the products they sell.
- *Emergent Social Networks*: In the age of almost-instant, wide information dissemination, feedback loops of influence among highly fashion-conscious groups of people may result in rapid evolution and exposure of styles.

5.3 Modeling Obligations

The subjective influence network model simply lays a framework for building an ontology that is capable of representing some aspects of subjectivity in fashion. In order for this model to be practically useful, a full ontology would need to be constructed, including:

- Enumerating (at least some of) the members in G, Φ, N, and M.
- Characterizing a relevant set of subjective functions $s()$.
- Calculating, estimating, or assuming values for the quads (t, i, m, a) for the edges between the nodes $x \in N$.
- Consideration of cycles. For example, 70's Disco fashion has come back in multiple times in past decades. The approach described here would model this return as a new style that is heavily influenced by the original. However, explicit modeling of this dynamic would be important.

5.4 First Order Interpretations of the Network

Interpreting an existing fashion network might allow us to make useful, testable judgements, including identification of important styles properties, including:

- *Novelty*: This is the subjective notion of a style that is different from previous styles in a pleasantly surprising way. Using our influence network model, one naive first order measure of a style's novelty is that the sum of intensity of influencing styles is low; i.e. that it is influenced only weakly by the combination of all previous styles. The novelty v of y could be defined as:

$$v_y = e^{-\sum_{x \neq y \in N} i_{xy}} \tag{2}$$

  where N is our influence network, and i_{xy} is the intensity element of $\overrightarrow{\mu_{xy}}$. In this case, when the sum of i values is high, $v_y \approx 0$, and when i is zero, $v_y = 1$. Other, more sophisticated measures of novelty could include deeper network analysis approaches or more nuances summing of intensity based on mechanism m and or agent a. This measure of novelty requires there to be no missing nodes or links in the influence network. A more sophisticated variant that tolerates missing information and noise would likely be needed in a practical application.

- *Impact*: This is a measure of how much a particular style has influenced all other styles as a whole. A simple (and very naive) measure of impact ι of a style x on the network N could be:

$$\iota_x = \sum_{y \neq x \in N} i_{xy} \tag{3}$$

If x has little impact then $\iota_x \approx 0$ and if x is heavily influential, then ι_x would be large. There is much previous, mature work on the topic of measuring influence in networks such the concepts of centrality, node influence metrics, page rank, etc. As such is beyond the scope here, and a likely important direction for future research.

6 Evaluation Approaches

In this section, we propose quantitative evaluation strategies to assess the practical usefulness of representing knowledge in the fashion domain using the influence network model presented here. In particular, we suggest measurements on values of such representation and potential applications.

6.1 Quality Measurements

In order to assess practical values of the proposed approach, we describe a number of evaluation strategies to measure its quality. Specifically, we focus on data-driven and task-driven evaluations which have been applied to ontologies in other domains (Raimond and Sandler 2012). For the former, we aim to measure how well the ontology represents empirical data related to fashion. For the latter, we examine information retrieval and recommender systems which could be consumers of the ontology and data.

Domain data approximation This is a data-driven approach to quantify how well the proposed ontology approximates empirical data in the fashion domain. Since fashion is a highly non-static, subjective and high-dimensional domain, we propose a few metrics which may capture expressiveness, both in terms of topics and temporal evolution, including:

- **Categorical precision**: Count how many styles encoded in the influence network are real-world recognizable styles in empirical domain data.
- **Temporal bias**: If we repeat the above categorical measurements on datasets from different time spans, the resulting metrics might stay stationary if this property of influence network's is time-invariant; otherwise, a network which fails to represent future datasets could indicate variable predictive power. The length of time span before the divergence is the representativeness timescale of the network, and the scope indicates its robustness.
- **Semantic similarity**: This measurement provides a distance metric between a traditional ontology augmented with an influence network and the text data in fashion domain in terms of "meanings" they express. We project both labels (class names O_c and property names O_p) in the ontology and the tokens in the text corpus D in the same vector space (e.g., using word2vec). Then we compute the overall similarity based on all labels' distances weighted by their importance within the network:

$$\sum_{c_i \in O_c} \sum_{t \in c_i} Sim(t, D) * \theta(c_i) + \sum_{p_i \in O_p} \sum_{t \in p_i} Sim(t, D) \qquad (4)$$

The importance score $\theta(c_i)$ is a normalized score $[0, 1]$ and can be defined depending on the context and usage. For example, $\theta(c_i)$ refers to how knowledgeable a network is w.r.t. class c_i, which can be approximated by the cumulative distribution function of its number of attributes u_i

$$\theta(c_i) = \sum_{u_m < u_i} P(U = u_m) \qquad (5)$$

Task-specific expressiveness This is a task-driven measurements to quantify how expressible the ontology is compared to user's mental representation in the context of a task. Here we use information retrieval in the fashion domain as an example task, and developed statistical measures of "expressiveness".

- **Query concept recall**: Given a query stream like "natural fabric button down from banana republic", we derive a mapping between concepts that appeared in the query and concepts encoded in the ontology. Specifically, we ask expert judges to identify important classes $\{c_i\}$ in the query. For each class being detected, we ask them to map it to the most similar label in the ontology. Then we compute the recall of concepts in the query as

$$\frac{\#\text{concepts mapped to ontology}}{\#\text{concepts detected}}$$

- **Search result ranking**: We use rankings of search results for a given query as a proxy of "golden standards". Then the Normalized Discounted Cumulative Gain (NDCG) metric can be adapted to measure the ontology's relevancy to search quality. Specifically, the "gain" is quantified by the ontology's recall of concepts in search results. We examine the top K returned search results. For each of them D_p at position p, we compute the ontology's document concept recall, which is then discounted by its logarithmic rank $log(p)$.

$$\sum_{p=1}^{K} \frac{Recall(O, D_p)}{log(p)} \tag{6}$$

7 Fashion Knowledge Graph Applications Using Subjective Influence Networks

7.1 Web Data Markup Through Schema.org

There are massive data sources on the Web (including mobile applications). However, they are mostly unstructured, and there is no common vocabulary which facilitates collective curation of domain knowledge. Schema.org markup has been the major adoption for web data (about 31.3% of all pages by Dec 2015 (Guha et al. 2016)) and is used by a variety of high traffic applications like search engines and news portal. Although it contains vertical specific schemas such as movies, music, medical and products, schema that can represent fashion content is absent.

In Fig. 6 we illustrate that the proposed ontology framework can easily adapt to a lightweight ontology and integrated as an external extension of the core schema.org vocabulary, while also linking to other relevant common vocabularies such as the *GoodRelations* for E-commerce (Hepp 2008), *SIOC* for influence mechanisms on the social Web (Breslin et al. 2005).

Consumed by machine learning systems The data represented in a subjective influence network as proposed here could be used for a variety of different data analysis and processing efforts, including the following types:

- **General machine learning problems**: The knowledge base represented by a populated influence network would contain instances associated with high-quality categorical types, which provide labeled data to train models for entity recognition. Also, the edges on their own in the network contain both numeric and categorical features which can be used in whole-network modeling experiments.

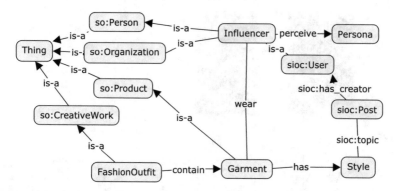

Fig. 6 Integration with schema.org

- **Fashion data retrieval**: The integration with http://schema.org enables community content publishers to explicitly annotate their posts with their perceived subjectivity of fashion contents, which are basic building blocks of a crowd-sourcing system. As a result, the marked up Web data in return allows for information organizers such as search engines to index rich contents and answer queries which contain both entities and subjective projections.
- **Recommender systems**: The taxonomy defined in the ontology provides a perfect complement to recommendations learnt in a bottom-up fashion. Therefore, it could be a very useful approach to deal with data sparsity situations such as cold start problems.

We now propose some concrete different use cases where the proposed work (Bollacker et al. 2016), can be applied.

Example 1 **Virality in Fashion** We may want to predict what new garments just seen on Milan Fashion week are going to become viral, to better supply our warehouses in the next 5 years (see Fig. 7).

- **Question 1**: Does fashion spread in the same way as viral music/art/cinema influence mechanisms function?
- **Question 2**: Can we use game virality mechanisms (e.g. Pokemon Go) to predict fashion virality?
- **Question 3**: Can a musical influence network be used to construct a predictive fashion influence network and find closely influential styles? Several hypothesis will have different graph instantiations in reality with different vectors of influence.
- **Hypothesis 1**: Social media mechanisms such as Facebook and Instagram are influential agents' tools of spread.
- **Hypothesis 2**: Counterfeiting, as another explicit mechanism of influence (copy) from the Western word.
- **Experiment**: Collect social media popular mobile games (Pokemon Go), and popular fashion items (bare shoulder blouse) and see time elapsed since first players/wearers have it until the game/garment style becomes viral. If in Asian countries

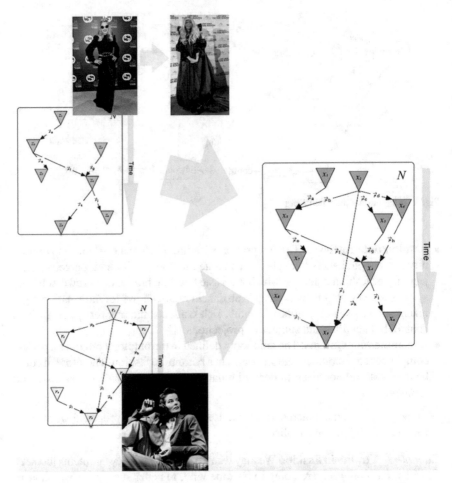

Fig. 7 Use case 1: **Mapping Network**s: Can influences in other cultural domains be used to predict fashion influences?

such as China, there are samples of the game/style in similar proportions relative to other games/garments within the same percentage rates to what it occurs in the same time span in the Western world, we can confirm our hypothesis 1. On the contrary, if there are not similar percentages of presence of "viral games" /"viral garment styles" in the Asian countries, we can contradict our initial hypothesis of Instagram being the agent. In the latter case, there must have been another agent conducting the influence (since Facebook and Instagram are banned in China), and further copyright experiments can be done to confirm hypothesis 2.

Example 2 Quantifying subjective attributes such as *retro* glasses. We want to build a quantitative model of what "retro" means, e.g., to distinguish it from "classic" , or

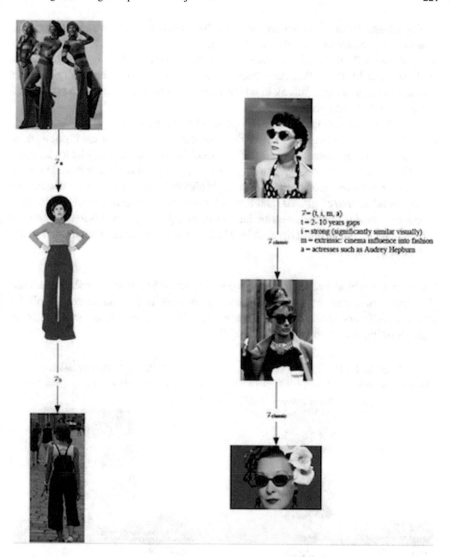

Fig. 8 Use case 2: **Quantifying Subjective Attributes**: Can influence networks characterize the differences between *retro* and *classic* glasses and their mechanisms of influence?

for this term to be computable, retrievable or indexable in our fashion Knowledge Graph advanced search engine (see Fig. 8).

- **Question 1**: How to characterize quantitatively the attributes of what retro glasses mean?
- **Question 2**: Do retro and classic glasses belong to the same kind of persona/ style/ physical attributes?

- **Hypothesis 1**: Retro glasses are those that have been worn since mid-century, they stop being in fashion, and they come back.
- **Hypothesis 2**: Retro glasses may or not have been carried by an icon. Retro glasses can be those that have been worn by a second degree family member (i.e., grandparent generation), in which case, the influence mechanism is based on appreciation/ homage.
- **Hypothesis 3**: Items that never really go out of fashion (i.e., classic Rayban glasses)
- **Experiment**: After representing each hypothesis in our ontological framework of influence, we can measure cyclicality between re-appearances of glasses in terms of their strength of presence within a given time window. If there is a clear time series strength based on social media celebrities wearing the different variations of "classical" or "retro" glasses, and that influence fades down in periods where they become non-fashionable, we can determine, based on a given seasonality strength parameter or visual similarity metric, which of the influence network models better fits our hypothesis.

Example 3 Fashion cyclicality: e.g., predicting when bell trousers will be fashionable again. Bell trousers and wide trousers have been in fashion on and off all the time with different variations. Could we predict when they will be in fashion strongly again, looking back at the past? (see Fig. 9).

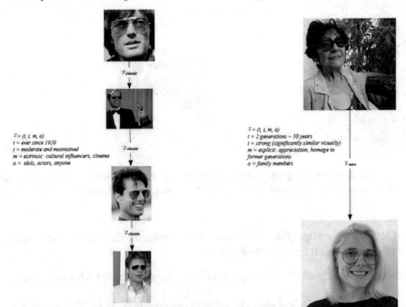

Fig. 9 Use case 3: **Predicting Fashion Cyclicality**: Can influence networks tell apart fashion cycles periodicity to predict when bell trousers will be fashionable again?

- **Question 1**: Is there a fixed interval which needs to happen in time for bell trousers to be out of fashion, for people to miss them again and they to come back?
- **Question 2**: Does every time that wide trousers became fashionable, they do it with a nuance or variation?
- **Question 3**: Is there a cyclic period where the exact classical bell trousers will always come back and if so, is this period predictable?
- **Hypothesis 1**: A minimum of 10 years need to happen for people to forget they already got enough of bell trousers.
- **Experiment**: Collect images from different decades where it was rare to find people wearing bell trousers and periods where most of people were wearing them, and annotate their main 1-year wide intervals when there significantly different peaks on these two wearing patterns. Execution examples:

  - Collecting as much data points (images of bell trousers wearers) as possible by recovering old black and white pictures since Google Trends only works from 2004 onwards.
  - Perform deep learning to separate silhouette representations from other trousers style features such as colour or material to gather a large dataset of bell and not bell trousers for each decade.
  - Perform a t-test to find if there is statistically significant difference in between periods where bell trousers were fashionable versus not fashionable. If there isn't statistically significance among "bell trouser fashionable" periods, and we can repeat the experiment for min. 4 peaks of seasonality/decades we may conclude such cyclic behaviour occurs in time and be able to predict when the effect will occur next (better being able to prepare for manufacture and inventory optimization).

Broader examples integrating social media across channels, could validate hypotheses on the influence mechanism aspect (e.g., being the mechanism social media or counterfeiting). An example of ways to refute/validate the hypothesis on the mechanism assumed could be by correlating search terms with social media co-occurring hashtags (see Fig. 10).

These were examples where machine learning problems could benefit of using ontology augmented computational intelligence techniques. Experiments in other domains, such as human activity recognition (Díaz Rodríguez et al. 2014a, b), show that the coupling of data-driven approaches together with fuzzy ontologies versus the use of uniquely crisp (simple) ontologies, are successful approaches in computational modelling of real-life problems. The reason is that some domains require to handle uncertainty, vagueness, imprecision or missing data specifically accounting for context and in a rich expressive manner. Fuzzy ontologies are a good formalism candidate to model such domains, for instance in online matchmaking (Ragone et al. 2008).

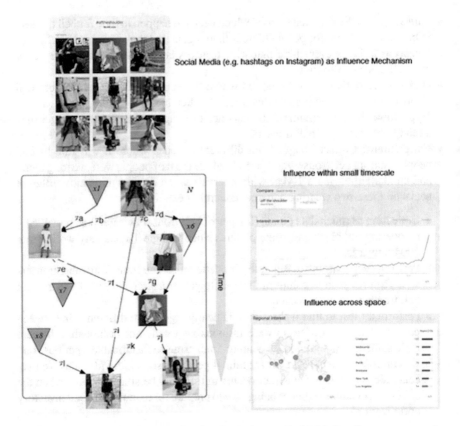

Fig. 10 Use case 4: **Representing Fashion Evolution on Social Media**: Can we measure the "viral" nature of rapid style changes?

8 Business Development Around Knowledge Graphs

A potential business development of tools built on top of the fashion knowledge graph, such as, e.g., style advice note writing assistant tools would be, for instance, an internal dashboard tool to aid decision making for stylists. Other than the in-house utility of such efficiency decision support tool, in this section potential business interests for e-commerce sites and personalization tools for commercializing B2B products are discussed. Clients also could benefit of this tool if it is robust enough, to avoid having the customer fill all the 50 question profile questionnaire (some customers, specially for men, do not finish completing).

8.1 Monetization of Fashion Knowledge Graphs

According to new reports, the virtual digital assistant market will reach $15.8 Billion by 2021.[11] Looking at a stylist note writing assistant, we can learn from similar models in the market on the fashion and also media analytics (both NLP and computer vision). How could a stylist note writing assistant help in other domains? Helping lowering the cognitive load of assembling an outfit or look within a *fix* box is the immediate benefit within SF organization. Estimating the benefits of the tool outside this concrete domain is difficult, since it is something that is not sold as a service and each domain would most probably require some fine-tuning. However, a B2B API can be provided for text analytics and insight extractions that can justify not the direct earnings or lift for the company (which may be difficult to directly measure/evaluate), but they may better justify the customer satisfaction, better improve the customers' requests and lead to less disatisfaction and churn. For instance, a tested campaign is the so-called "Want her back" (the possibility to get the same stylist to assemble your look next time if a customer thought the stylist got his/her style). Measuring the increase in the amount of customers wanting to stick to the same stylist is an implicit feedback metric worth considering as for customer experience improvement.

8.2 Business Cases on Fashion, NLP and Dialogue-Based Personalization

There are some existing services that aim at personalizing fashion in one or other ways: *Cladwell*[12] (provides advice and a small good quality *capsule* seasonal set that provides a minimal wardrobe with your favourite pieces every season), *Bombfell* (the male version of Stitch Fix), *Trunk Club* (a higher price point version), and *Lookiero*[13] (Spanish version of Stitch Fix) are similar services to Stitch Fix that also use fashion personalization and delivery. In the clothing delivery + renting space, *Rent the runway* and *Le Tote*[14] are similar providers but for second hand clothing. *Boon + Gable*,[15] on the other side, include in the service the visit of the personal stylist with clothes to your home for you to get in-person advice and combine with your existing wardrobe. However, Stitch Fix is the oldest service, with 5 years of experience learning from data, and where personalization and human in the loop are key competitive advantages that make use of machine learning and internal crowd-sourcing tools.

---

[11] https://www.tractica.com/newsroom/press-releases/the-virtual-digital-assistant-market-will-reach-15-8-billion-worldwide-by-2021/.

[12] https://cladwell.com/.

[13] https://lookiero.com/.

[14] https://www.letote.com/.

[15] https://www.boonandgable.com/.

On the information extraction and information extraction domain, some more general machine learning NLP tools as a service exists, such as *nuun.io*, *FoxType*[16] (helps detecting politeness among others), or Receptiviti[17] (psychology insights for analyzing tone and other aspects in natural language). Based on text-based interfaces, the bloom of chat bots is unstoppable (in Facebook Messenger, Slack, etc.); however, these are merely for customer service (helloaida.io) or informative (e.g., Immigration attorney *visabot.co*, *Polly.datalog.ai* and *getAsteria.com* for conversational/health/personalization purposes).

Other models having NLP analytics as the core value proposition are writing assistants such as *Grammarly*,[18] *Boomerang*[19] or bot kits/conversational agents[20]

Business models to consider for such developed tool are also in the NLP space: *claralabs*, *x.ai* or *Talla*[21] are core NLP based meeting assistants. *Olivia.ai*, *MoneyStrands.com* and *Bond*[22] are finance personal assistants, but there is also AI chat assistants for shopping such as *kip*.[23] In the fashion annotation computer vision scene, examples such as *wide-eyes.it*, *21buttons.com* or *chicisimo.com* offer the counterpart to the text only annotation tools.

One time service could be as well of use for real-time outfit assessment, taking a model of a consultation fee, such as in *Remedy* remote health care assistance.[24]

8.3 Case Study Survey: Potential Business Proposition on Digital Style Advisors

A small survey done in social media shows the potential market of such one-time transaction style advice service. Results in Figs. 11 and 12 show that 34% of people would be willing to pay for having some kind of last minute style advice based upon a *selfie* before leaving home, while 66% would not. This demonstrates that special occasions must be carefully studied and despite being a not large margin of potential benefit, it could be of business value (e.g. getting tips on mixing and matching garments for an interview, meeting, date, ceremony or other special events). Usage

---

[16]https://foxtype.com/.

[17]http://www.receptiviti.ai/.

[18]http://www.grammarly.com.

[19]http://www.boomerangmail.com/respondable/?utm_medium=email& utm_source=user+email&utm_campaign=respondable+launch&utm_content=bottle+banner.

[20]https://howdy.ai/botkit/, https://rundexter.com.

[21]http://talla.com/office-management.

[22]https://medium.com/@uday_akkaraju/meet-bond-a-bot-thatll-make-you-richer-3edd2540bf06#.2mdudz88g.

[23]https://www.kipthis.com/.

[24]Remedy costs $15 per consultation done by professionals remotely, instead of avg. $80. https://www.remedymedical.com/.

NataliaDíazRodríguez
@NataliaDiazRodr

Would you pay for a real-time outfit selfi assessment before leaving home to make sure you are wearing your best and most stylish look?

24% Yes, only if AI-powered

3% Only by fashionistas

7% Yes, if human+AI powered

66% No

29 votes · Final results

RETWEETS 3   LIKES 3

11:44 AM - 15 Dec 2016 from Fuenlabrada, Spain

Fig. 11 Social media survey results for potential monetization

NataliaDíazRodríguez @NataliaDiazRodr
Would you pay for a real-time outfit selfi assessment before leaving home to make sure you are wearing your best and most stylish look?

Reach a bigger audience
Get more engagements by promoting this Tweet!

Get started

Impressions	766
Total engagements	41
Votes	24
Detail expands	7
Retweets	3
Likes	3
Profile clicks	3
Replies	1

Fig. 12 Potential monetization for viability of business models: impressions

of existing platforms such as *Snapchat* or *Instagram*, which allow video social sharing immediately, are proposed as areas to explore.

Would you pay for a real-time outfit selfie assessment before leaving home to make sure you are wearing your best and most stylish look? This is the question to evaluate if a style advice agent would be worth building and if so, how much would people pay for it.

The transaction to pay for could be per picture asking advice for. Finding interest in the product idea involves having a prototype beta test tool providing advice from curated stylists with a karma and reputation feedback-based system. For stylists, it would let them build their reputation and grow their influence audience. This could also be, as a side effect, a way for companies like Stitch Fix to recruit stylists and augment the style advice knowledge base *bible*. The final product can be an app where the user send their closet pictures, or a picture of their look just before going out to that job interview, your fourth date, or your conference talk. Such a product can only be built after data showing interest is gathered. Key questions are: Are consumers really willing to pay for this service? How to guarantee there is going there a stylist there when you need it? How will the expert advice quality control be executed?

Because of these issues, the aim of this survey was to strictly find if users would be willing to pay for, e.g. an AI/chatbot-like service giving professional, *fashionista*, or peer advice on the go.

The Twitter survey reached 766 impressions, 41 total engagements, 24 votes, 7 detail expands, 3 retweets, 3 likes, 2 profile clicks and 1 reply in only 24h (see Figs. 11, and 12). A next step would be finding what are the minimum requirements the users would want the *fashionista-bot* to have. The same survey conducted in Facebook without choice answers (i.e., open question) resulted in a larger proportion of negative comments such as the following (combined with 11 *likes*):

- *No, because there are more important matters in this world! ;)*
- *Yes, because what you wear is a perspective of your taste!*
- *Then why would you let anybody else decide on it, instead of yourself ?*
- *You cannot say that what you are now wearing is only decided by you!! before even you want to buy a piece of clothing, it is already decided by designers and fashion gurus. ;) you only pick and try to come up with a set of outfit. I would rather have a professional/intellectual idea on that before wear something hideous!*
- *No, I do not even use it for free.*
- *Nope, a mirror is enough*
- *I would pay not to get assessed*
- *No*
- *Personally I would not like anybody to "tell me" what I should be wearing. I like, though, an app that would suggest an outfit in the morning. I think there could be some use cases for this though: people that need to look professional each day but have no time or interest to decide their own outfits every time. In that case the app could be your "personal stylist", although I'm uncertain of how good the app could decide if something flatters you or not, because its not only a matter of proportions... If the clothes and outfits would already be "verified" by a stylist beforehand, then the risk of miscalculations gets smaller. Also, in the case that the app really would know what looks good on you or not, what are the chances of it being "creative"? A lot of inventions, art and fashion arises from accidents, miscalculations or just wild fantasy, so there is always the risk of the app being too safe and boring :) But interesting thought anyhow! I'm always interested in questions of fashion, although i prefer to ponder on them without a computer normally :).*
- *Yes! This kind of app would be my hero since I do not need to think about my outfit of the day. And believe me sometimes it takes half an hour to come up with a nice outfit! Personally speaking, I have a strong connection to my outfit :D It can change my mood easily and when I am not wearing based on my standards, I feel uncomfortable in my skin. I really do not care what others think about me! I care*

what I feel about myself! Even more exciting option would be to have an app to suggest your outfit in the morning! Sounds cool!
- *The correct attire is always combat boots, cargo pants and a black tee-shirt. Add sunglasses and dog tags if you're feeling fancy. Regardless of gender. :)*
- *I would not.I think it's time consuming, i don't trust apps to do this yet... Maybe in 20 years.*
- *Thanks but no thanks*
- *I would use the money and buy a stylish brand!*
- *No*
- *Nope. I wouldn't want to be paid for having it. My outfit is my decision. How does this thing not sound like society telling you how you should dress? (especially for women I guess)*
- *Nope*

A "magic mirror" application for a last minute"Am I looking great?" before exiting home could be interesting, but according to the survey results, it is of doubtful value, since it appears to exist a very niche and limited prospect on acquiring paying users. The survey would require a final tool prototype for further exploration and business development refinement.

8.4   Target Market

Customers of machine learning-targeted media (text and image) annotation tools can serve a B2B model for customers of a varied range. The range of potential customers includes a very diverse range of business, considering the fashion world only initially, for instance, customers of subjective influence networks could include:

1. *Chicfy.com* A service for online second hand sale and swap. The system's business model consists of taking a commission for each transaction made.
2. *Vinted.com*: An application similar to the previous system with no commissions.
3. *21Buttons.com* A service for users to rate other users outfits and find and buy with one click the clothes worn. Users uploading an outlook picture get a commission if other users buy any of their exhibited items.
4. *wide-eyes.it* A service that provides widgets for online shops to show visually similar elements to the searched ones.
5. *CrowdStar*[25] A fashion mobile game where users dress digital mannequins and receive credits if their looks have the highest ranking in n. of votes. The business model is in-game purchases (for a more varied wardrobe), clothes are real and can be bought.
6. *Betabrand*[26]: A site where anyone can create and upload any garment design, crowd-fund or crowd-source garment patterns. Betabrand uses "Crowd Funding Predictive Modeling" to achieve market research through crowd funding and help

---

[25] www.crowdstar.com/.

[26] www.betabrand.com/.

in accurately predicting inventory adoption rates and future demand of catalogues to minimize the risk of a poorly-performing product or excess inventory.

7. Media data-driven insight providers, such as Secret Sauce Partners,[27] a Budapest and San Francisco-based company that offers a data-driven merchandising (DDM) platform for *fit predictor* (finds a shopper's best fitting size in seconds), a *style finder* (apparel shopping by visual features) and a *outfit maker* (automated outfit recommendations based on stylistic matches).[28]

8. *E-commerce sites* that are not fully tech and data-driven, such as *Gilt* or *Stylect*[29] or *therealreal.com* (luxury consignment sales), and *Best Secret*[30] (invitation only luxury discounted sales).

9   Conclusions

In this paper we propose a new ontological augmentation in the fashion domain, which represents subjective feature information as an influence network. Because fashion (just like art, music or languages) strongly contains subjective information (cultural phenomena which are not designed nor engineered), we believe that such an augmentation might result in the construction of higher performing machine learning and data analysis systems.

Following the theoretical modeling, we suggest quantitative measures to assess the framework's utility to machine learning systems. Especially we focused on quantifying how well the ontology can represent domain data, and how the features from an influence network could be integrated into machine learning systems.

9.1   Future Work

Future work on cognitive cities use cases and applications within the culture, fashion or leisure domains should focus on ubiquitous but effortless recommendation systems based on the user's digital and off-line footprint. In fashion, an example along these lines is the aim of the application that results from H&M's partnership with Google for creating a customized "data dress" based on the places a person hangs out after work or the restaurants she visits.[31] Other example of fashion trend spotting on Google is using multiple markets focusing on apparel trends to enable a better understanding of how trends spread and behaviors emerge across markets.[32] Other ideas can extend

---

[27] www.secretsaucepartners.com/.

[28] http://tech.gilt.com/2015/01/27/new-gilt-product-feature-fit-predictor.

[29] https://www.gilt.com/, http://tech.gilt.com/, www.stylect.com.

[30] www.bestsecret.com.

[31] http://tinyurl.com/hxg6r7e.

[32] https://storage.googleapis.com/think/docs/twg-fashion-trends-2016.pdf.   http://www.businessinsider.com/google-partners-with-hm-ivyrevel-for-coded-couture-project-2017-2 http://www.ivyrevel.com/.

the work of (Vaccaro et al. 2016) as well as our illustrated examples (poster[33]). Interestingly, the work in (Vaccaro et al. 2016) uses translational topic coherence among some crowd-sourcing tasks in order to translate (abstract) style-language into (concrete) element-language and, in this way, generate recommendations from natural language requests or subjective description notes. This is a step that, against Stitch Fix's philosophy, has as target to achieve a fully digital stylist (without the human in the loop). In this work, although PLTM (Polylingual topic model) were not originally intended to support direct translation between languages, in domains where word order is unimportant, given a document in one language, they can be used to produce an equivalent document in a different *language* by identifying high probability words.[34]

However, the extraction of the most precious information from experts in a context-aware manner and its continuous integration in a broad dimensionality space for a timely recommendation is still a large part of the machine learning bottleneck. These areas are those where future resources can be well allocated in machine learning in general.

Future work on subjective influence networks will instantiate concrete machine learning problems into the proposed approach. For instance, an example can be quantifying fuzzy influence networks in social media opinions (Wang and Mendel 2016). In this way we will validate our theoretical assumptions by incarnating and materializing different influence functions, distance and quality measures, scales and other parameters for our model assessment and evaluation in different machine learning problems. Future efforts should also be put into considering the integration of KGs into black box neural recommendation pipelines.

Integration of both text and images would be another area to enrich collective intelligence and a way of achieving data pooling for context awareness. For instance, *wide-eyes.it* focuses on producing visually similar images on a B2B manner. Integrating insights of text together with images is paramount for cross-referencing. Embeddings methods integrating both text, image and social influence are still in early stage. More automatic ways of integrating expert feedback and new approaches to having the human in the machine learning loop should be explored in the area of unsupervised reinforcement learning (Abbeel and Ng 2004; Thomaz and Breazeal 2006). Exploiting other social network channels' text feeds is also an under-explored area within most organizations (which are limited to analyze only the one or two most popular social feeds). Only in this way we will be able to assess properly reward functions that affect and evaluate the system's learning as a whole.

Ultimately, these actions will help refining the model's capability to effectively quantify influence and subjectivity in fashion, style and other subjective and more volatile domains.

---

[33] https://github.com/NataliaDiaz/PostersAndPresentations/blob/master/posters.

[34] E.g., given a document in the element language, we can infer the topic distribution for that document under the trained model: *Since the topic distribution for a document will be the same in style language, we can produce an equivalent outfit in the style language by identifying high probability words in that language* (Vaccaro et al. 2016).

Acknowledgements This work was done within an internship program of the EIT Digital doctoral school from EU as part of the Innovation & Entrepreneurship (I&E) curriculum. We thank EIT Digital Doctoral School, advisors Prof. Johan Lilius and PhD. Jussi Autere for the financial and human support throughout the Innovation & Entrepreneurship education. From Stitch Fix team we thank Jay B. Martin for his generous editorial assistance and management, Ian Horn for mentoring, and the rest of the Human Computation and Algorithm and Analytics teams at Stitch Fix for the inspiring and collaborative atmosphere (John McDonnell, Xian Li, Katherine Livins, John Clevenger, Roberto Sanchís Ojeda, Hoda Eydgahi, Jay Wang, Akshay Wadia, Sky Jin, Sonya Berg, etc.). Likewise, we thank the EU COST Action on.

References

Abbeel P, Ng AY (2004) Apprenticeship learning via inverse reinforcement learning. In: Proceedings of the twenty-first international conference on machine learning, ACM, p 1

Ajmani S, Ghosh H, Mallik A, Chaudhury S (2013) An ontology based personalized garment recommendation system. In: 2013 IEEE/WIC/ACM International joint conferences on Web Intelligence (WI) and Intelligent Agent Technologies (IAT), vol 3. IEEE, pp 17–20

Bollacker K, Díaz-Rodríguez N, Li X (2016) Beyond clothing ontologies: modeling fashion with subjective influence networks. In: KDD workshop on machine learning meets fashion

Bollacker K, Evans C, Paritosh P, Sturge T, Taylor J (2008) Freebase: a collaboratively created graph database for structuring human knowledge. In: Proceedings of the 2008 ACM SIGMOD international conference on management of data, ACM, New York, NY, USA, SIGMOD '08, pp 1247–1250. https://doi.org/10.1145/1376616.1376746

Breslin JG, Harth A, Bojars U, Decker S (2005) Towards semantically-interlinked online communities. In: The semantic web: research and applications, Springer, pp 500–514

Díaz Rodríguez N, Cuéllar M, Lilius J, Calvo-Flores MD (2014a) A fuzzy ontology for semantic modeling and recognition of human behavior. Knowl Based Syst 66:46–60

Díaz-Rodríguez N, Cadahía OL, Cuéllar MP, Lilius J, Calvo-Flores MD (2014b) Handling real-world context awareness, uncertainty and vagueness in real-time human activity tracking and recognition with a fuzzy ontology-based hybrid method. Sensors 14(10):18,131–18,171. https://doi.org/10.3390/s141018131

Díaz-Rodríguez N, Cuéllar MP, Lilius J, Calvo-Flores MD (2014c) A survey on ontologies for human behavior recognition. ACM Comput Surv 46(4):43:1–43:33. https://doi.org/10.1145/2523819

D'Onofrio S, Portmann E (2017) Cognitive computing in smart cities. Inform-Spektrum 40(1):46–57. https://doi.org/10.1007/s00287-016-1006-1

Hepp M (2008) Goodrelations: an ontology for describing products and services offers on the web. In: Knowledge Engineering: Practice and Patterns, Springer, pp 329–346

Li J, Li Y (2012) Cognitive model based fashion style decision making. Expert Syst Appl 39(5):4972–4977. https://doi.org/10.1016/j.eswa.2011.10.017

Wang L, Zeng X, Koehl L, Chen Y (2015) Intelligent fashion recommender system: Fuzzy logic in personalized garment design. IEEE Trans Hum-Mach Syst 45(1):95–109

Miller J (2011) Fashion and music. Bloomsbury Publishing, https://books.google.com/books?id=F7XKJkhkRCsC

Novalija I, Leban G (2013) Applying NLP for building domain ontology: fashion collection, pp 147–150

Ragone A, Straccia U, Bobillo F, Di Noia T, Di Sciascio E, Donini FM (2008) Fuzzy description logics for bilateral matchmaking in e-marketplaces. In: Description logics

Raimond Y, Abdallah SA, Sandler MB, Giasson F (2007) The music ontology. In: ISMIR

Raimond Y, Sandler M (2012) Evaluation of the music ontology framework. In: The semantic web: research and applications, Springer, pp 255–269

Guha R, Brickley D, Macbeth S (2016) Schema. org: Evolution of structured data on the web. Commun ACM 59(2):44–51

Scott AJ (2014) Beyond the creative city: cognitive-cultural capitalism and the new urbanism. Reg Stud 48(4):565–578. https://doi.org/10.1080/00343404.2014.891010

Steinfeld A, Bennett SR, Cunningham K, Lahut M, Quinones PA, Wexler D, Siewiorek D, Hayes J, Cohen P, Fitzgerald J, Hansson O, Pool M, Drummond M (2007) Evaluation of an integrated multi-task machine learning system with humans in the loop. In: NIST performance metrics for intelligent systems workshop (PerMIS)

Thomaz AL, Breazeal C (2006) Reinforcement learning with human teachers: evidence of feedback and guidance with implications for learning performance

Vaccaro K, Shivakumar S, Ding Z, Karahalios K, Kumar R (2016) The elements of fashion style. In: Proceedings of the 29th annual symposium on user interface software and technology, ACM, New York, NY, USA, UIST '16, pp 777–785. https://doi.org/10.1145/2984511.2984573

Vogiatzis D, Pierrakos D, Paliouras G, Jenkyn-Jones S, Possen BJHHA (2012) Expert and community based style advice. Expert Syst Appl 39(12):10,647–10,655. https://doi.org/10.1016/j.eswa.2012.02.178

Wah C, Branson S, Perona P, Belongie S (2011) Multiclass recognition and part localization with humans in the loop. In: 2011 IEEE international conference on computer vision (ICCV), IEEE, pp 2524–2531

Wang L, Mendel JM (2016) Fuzzy opinion networks: A mathematical framework for the evolution of opinions and their uncertainties across social networks. CoRR, http://arxiv.org/abs/1602.06508

Zoghbi S, Heyman G, Gomez JC, Moens MF (2016) Cross-modal fashion search. In: MultiMedia Modeling, Springer International Publishing, pp 367–373

Zoghbi S, Vulić I, Moens MF (2013) I pinned it. where can I buy one like it?: Automatically linking pinterest pins to online webshops. In: Proceedings of the 2013 workshop on Data-driven user behavioral modelling and mining from social media, ACM, pp 9–12

Kurt Bollacker is computer scientist who has published in the areas of machine learning, digital libraries, semantic networks, and electrocardiographic modeling. His research interests include graph databases, long term data preservation, and human/machine collaborative systems. He is currently Digital Research Director of the Long Now Foundation and builds tools to enable scalable collaboration among data scientists at Stitch Fix, Inc.

Natalia Díaz-Rodríguez is Computer Engineer at the University of Granada (UGR), Spain, and holds a Double PhD from Abo Akademi, Finland (together with UGR) in Artificial Intelligence and Semantic Fuzzy Modelling for Human Behaviour Recognition in Smart Spaces. She worked on R&D at CERN (Switzerland), Philips Research (Netherlands) at the Personal Health Department, was postdoctoral researcher at the University of California in Santa Cruz and worked in the Silicon Valley at Stitch Fix (San Francisco, CA), a recommendation service for fashion delivery with humans in the machine learning-loop. Natalia D?az Rodr?guez participated in a range of international projects and is member of the Management Committee of EU COST (European Cooperation in Science and Technology), Action AAPELE.EU (Algorithms, Architectures and Platforms for Enhanced Living Environments). She was Google Anita Borg Scholar 2014, Heidelberg Laureate Forum (2014) and fellow (2017), and obtained the Nokia Scholarship. Currently, she is researcher and lecturer at ENSTA ParisTech at the Autonomous Systems and Robotics (computer vision) Group and INRIA Flowers team, France. She is working on deep and reinforcement learning for state representation learning within the robotics DREAM project.

A Dynamic Route Planning Prototype for Cognitive Cities

**Patrick Kaltenrieder, Jorge Parra, Thomas Krebs,
Noémie Zurlinden, Edy Portmann and Thomas Myrach**

Abstract A software prototype for dynamic route planning in the travel industry for cognitive cities is presented in this paper. In contrast to existing tools, the prototype enhances the travel experience (i.e., sightseeing) by allowing additional flexibility to the user. The theoretical background of the paper strengthens the understanding of the introduced concepts (e.g., cognitive cities, fuzzy logic, graph databases) to conceive the presented prototype. The prototype applies an instantiation and enhancement of the graph database Neo4j. For didactical reasons and to strengthen the understanding of this prototype a scenario applied to route planning in the city of Bern (Switzerland) is shown in the paper.

Keywords Cognitive city · Fuzzy logic · Graph databases · Neo4j
Software prototype · Travel industry

1 Introduction

Experts predict that the amount of people who live in cities will rise to 6.4 billion by 2050, which will be equal to 70% of the world's population (United Nations 2008). This increase makes it essential to find solutions to urban problems (e.g., transportation systems (D'Onofrio et al. 2016), power supply networks) and to ensure prosperous coexistence among citizens. The interdependent systems of these cities have to work properly to ensure a high quality of life for their citizens (Boisson de Marca 2015). This includes the capabilities to deliver enhanced touristic services for visitors. People visiting a foreign city are dependent on information to orientate themselves (e.g., finding accommodation, getting to know the transportation system, choosing restaurants, discovering sightseeing locations).

P. Kaltenrieder (✉) · J. Parra · T. Krebs · N. Zurlinden · T. Myrach
University of Bern, Bern, Switzerland
e-mail: patrick.kaltenrieder@imu.unibe.ch

E. Portmann
Human-IST Institute of the University of Fribourg, Fribourg, Switzerland

© Springer Nature Switzerland AG 2019   235
E. Portmann et al. (eds.), *Designing Cognitive Cities*, Studies in Systems,
Decision and Control 176, https://doi.org/10.1007/978-3-030-00317-3_10

Cities that are based on cognitive systems are called cognitive cities (Kelly and Hamm 2013; Mostashari et al. 2011). Cognitive computing and especially cognitive systems can make the proper functioning of cities possible, as they allow for better interaction between humans and computer systems. Traditional systems are not able to handle the scale, uncertainty, and complexity of data in a similar way (Hurwitz et al. 2015).

Interactions between humans and computer systems require systems to be able to handle the imprecision and uncertainty of human reasoning, perceptions, and their environment. Among other techniques, this can be achieved by using fuzzy set theory and fuzzy logic (Zadeh 1988). Fuzzy logic extends traditional, bivalent logic as it not only allows for the truth values of true and not true, but for anything in between (Zadeh 1965, 1988). Applying fuzzy logic, it is necessary to tackle the ambiguity and imprecision of reality and, thus, to enhance citizens lives (e.g., through adaptive bus schedules.) Perticone and Tabacchi (2016) presents possible computational intelligence approaches for cognitive city communication (e.g., metaheuristics, computing with words, fuzzy classifiers, fuzzy ontologies).

There already exist some applications with the aim of facilitating people's communication (e.g., Waze,[1] Google Now[2] and Snips[3]). However, there are few semi-automated tools that make interaction and organization easier for people (e.g., Sygic Travel[4]). There are some papers that made first attempts at developing such semi-automated tools, which present mobile applications based on the fuzzy analytical hierarchy process (FAHP) and fuzzy cognitive maps (FCM) to enhance communication and governance (Kaltenrieder et al. 2014, 2015a, b, 2016). The aim of this paper is to show another step that leads to cognitive cities by augmenting an existing graph database (i.e., Neo4j[5]) with fuzzy logic, based on fuzzy set theory, for a tool in the travel industry. Graph database management systems are able to model every kind of scenario, and thus to process queries from various fields, such as the natural sciences or social networks (Robinson et al. 2013; Jouilli and Vansteenberghe 2013; Miller 2013). The prototype presented in this paper will make it possible to process fuzzy queries with graph databases. Thus, numeric properties of edges (e.g., duration in minutes) between two nodes stored in a graph database can be represented as linguistic values (e.g., very short, long), which allows us to answer queries containing uncertainties. Because it is able to process imprecise data, this prototype can take advantage of big data, as it can handle its inaccuracies and ambiguities (i.e., big data's veracity Hurwitz et al. 2015; IBM 2015) through the scalability of graph databases.

A prototype is developed to apply the proposed idea to the travel industry, namely to the planning of sightseeing tours. Current tourist travel websites (e.g., MySwitzerland.ch[6]) mainly provide static routes that can be completed within a fixed time frame

[1]https://www.waze.com/.
[2]https://www.google.com/landing/now/.
[3]http://snips.net/.
[4]https://travel.sygic.com.
[5]http://neo4j.com/.
[6]https://www.myswitzerland.com/.

and whose points (i.e., starting point, sightseeing spots and end point) are fixed. To enhance the travel experience, it is important that a tool is adapted to the tourists needs and ways of thinking. Unlike computer systems, humans mostly think in vague terms and formulate it in seemingly fixed terms (i.e., "I would like to do a sightseeing tour for (about) three hours."). Computer systems have difficulties handling such imprecise data and processing fuzzy queries. Taking into account of fuzzy logic, the proposed prototype is able to overcome this shortcoming. The prototype presented in this paper provides flexible routes that can be changed according individual preferences (e.g., starting and end points, sightseeing spots, and the available time frame) by introducing cognitive computing. This will provide tourists with a large advantage, as they receive suggestions for tours that fit their needs based on fuzzy criteria. The prototype makes it possible to translate numeric values into linguistic terms or to assign a linguistic term to a range of numeric values. Thus, time or distance, for example, can be represented in linguistic (i.e., fuzzy) terms. This makes it possible for the underlying cognitive system to process fuzzy queries. Users of a tool based on this prototype are no longer faced with a limited amount of predetermined suggestions that very likely do not suit their particular needs. They can rather choose from options that are precisely tailored to their preferences, as the starting and end point and the amount of time available are determining variables in (cognitively) computing the sightseeing tours (Hurwitz et al. 2015; Kelly and Hamm 2013). The user can input the relevant information (i.e., starting and end points, sightseeing spots, and available time frame), and the personalized suggestions for different tours are visualized linguistically (e.g., tour one "good fit", tour two "ok fit") on the computer screen. In our opinion, this tool would represent a further step in the direction of cognitive cities, where the cities' systems and the citizens cooperate to enhance living together and welfare (Hurwitz et al. 2015).

The remainder of this paper is structured as follows: Sect. 2 presents the theoretical background that is crucial for the development of the proposed idea and the prototype (i.e., cognitive cities, fuzzy set theory and fuzzy logic, and graph databases). In Sect. 3 the prototype is presented. To illustrate a possible tool and the advantages of the prototype, a scenario is shown in this section. Section 4 concludes the paper and gives an outlook on further research.

2   Theoretical Background

This section gives an overview of the theories that build the basis for the proposed prototype. First, the concept of cognitive cities is presented in Sect. 2.1. Then, in Sect. 2.2, the characteristics of human perceptions and reasoning (i.e., partiality and granularity) are described, which allow us to reduce information and facilitate problem solving. As the imprecision of human perceptions and reasoning make the interaction between humans and computer systems difficult, theories have been developed to overcome these challenges (e.g., fuzzy set theory and fuzzy logic), which will be presented in Sect. 2.3. The prototype presented in this paper is based on these theories and thus allows for enhanced interaction between humans and computer systems. The prototype is developed to complement graph databases, which are considered to be convenient tools for data management and thereby enable their use for cognitive cities. An introduction to graph databases and different types are presented in Sect. 2.4. including an evaluation of our introduced graph databases.

2.1 Cognitive Cities

The term cognitive city was first used by Mostashari et al. (2011) and is an extension of the concept of smart cities by including cognition. A smart city enables sustainable growth and a high standard of living through investments in communication infrastructure as well as in human and social capital (Perticone and Tabacchi 2016; Caragliu et al. 2011). Smart cities are constantly improving their interaction with citizens by using technology. Kaltenrieder et al. (2014, 2015a) perceive cognitive cities as an extension of smart cities by including connectivism, a type of learning and cognition theory. Connectivism states that people are no longer able to learn based only on their own experiences, but have to rely on experiences of others because the knowledge base increases constantly (e.g., exploiting big urban data). This makes interactions with other people and technology crucial for learning (Siemens 2005). This paper is based on Kaltenrieder et al.'s (2015b) understanding of cognitive computing, acting as an umbrella that includes connectivism as well as various other aspects.

Through the interaction between humans and computer systems, both can learn, and their collective intelligence increases. This concept is called intelligence amplification loop (Kaufmann et al. 2012; Portmann et al. 2012). This learning process is characterized by emergence, which refers to the fact that new and coherent properties can arise spontaneously bottom-up from the interaction of micro-level components of a system, not following a certain plan (Goldstein 1999; Kaufmann and Portmann 2015). Thus, emergence is an important foundation of cognitive cities.

2.2 Perception, Partiality and Granularity

According to Zadeh (1999), there are many shortcomings[7] in our scientific and technological achievements (e.g., "we cannot build robots which can move with the agility of animals or humans; we cannot automate driving in heavy traffic; we cannot translate from one language to another at the level of a human interpreter […] and we cannot build machines that can compete with children in the performance of a wide variety of physical and cognitive tasks."). Zadeh sees the source of these limitations in the fact that we know ways to address measurements, but humans have not yet developed methods that would enable reasoning and computing with perceptions and

---

[7]Some of these shortcomings are yet to be overcome. Still, not in the granularity as intended by Zadeh.

propositions in natural language, as they are fuzzy and difficult to grasp. The human mind has the capability to exploit perceptions and thus to make rational decisions solely based on them, without using measurements or computation (Zadeh 1999). Perticone and Tabacchi (2016) present various computational intelligence approaches (e.g., metaheuristics, computing with words, fuzzy classifiers, fuzzy ontologies) to tackle these shortcomings.

To be more precise, there are two fundamental aspects of human cognition: partiality and granularity. Partiality means that concepts like understanding, truth, certainty, and similarity are not bivalent but a matter of degrees (Zadeh 2004). Granularity means the clumping and forming of granules (Zadeh 1997, 2004). Perceptions can be seen as fuzzy granules. A granule consists of several physical or mental objects that belong to each other because they are indistinguishable, similar, close to each other, or they have the same functionality.

Granules with sharp boundaries are called crisp granules, whereas granules with soft boundaries are called fuzzy granules. Soft boundaries mean that there is no abrupt change from membership to non-membership but rather a gradual change. Time can for example be granulated crisply into minutes, and fuzzily into intervals labelled very short, short, long, and very long. Thus, granulation means that an entity is divided into its parts (i.e., into a set of granules.) By applying fuzzy information granulation, humans are able to wrap up data, based on which they reason and make rational decisions in an imprecise and uncertain environment (Zadeh 1997, 1999). Information granulation (Zadeh 1979) allows to manage big data (Yao 2005), and thereby to reduce information (Lin 1997), which facilitates human problem solving because it allows gaining a first overview on relevant information (Yao 2000). The fuzziness inherent in human perceptions and reasoning make the interaction with computer systems difficult, as they cannot easily process imprecise information. Thus, a set theory and a theory of logic were developed, whose applications could enable computer systems to process fuzzy information and queries. These theories are presented in the next subsection.

2.3 *Fuzzy Set Theory and Fuzzy Logic*

Bivalent, traditional logic (also called Boolean logic) is not perfectly suitable for accounting for the imprecision and uncertainty of human environments, perceptions and reasoning. Thus, Zadeh developed a new theory of logic: fuzzy logic (Zadeh 1988), which is based on fuzzy set theory (Zadeh 1965). A fuzzy set is a generalization of a traditional set, which is also called crisp (Zadeh 1965) (see also Sect. 2.1). Elements do belong to a fuzzy set only to a certain degree. Thus, one element has a grade of membership between 0 (no membership) to 1 (full membership) in each fuzzy set. The membership function that defines a fuzzy set X in the space of objects $Z = \{z\}$ appears as follows:

$$h(X) = \phi,$$

where $\phi \in [0, 1]$,

the higher the value of φ, the higher the grade of membership of z in X (Zadeh 1965).

Often, linguistic variables are used as labels of the membership degree, (e.g., very short, long) (Zadeh 1988). Fuzzy logic, allows augmenting traditional logic by partial truth, where a proposition is not either true or false, but takes any truth value in between (i.e., varying degrees of truth are possible) (Zadeh 1988). Thus, computer systems can be enabled to process imprecise and uncertain information and fuzzy queries. The idea of the prototype presented in this paper is to enhance the (crisp) graph database Neo4j by integrating fuzzy logic.

The last two Subsections have presented the strengths of human cognition, the problems of the interaction between humans and computer systems, and how fuzzy logic can be augmented by fuzziness, which makes tools possible that allow computer systems to cope with reality's imprecisions and uncertainties. As the focus of this paper lies in the presentation of an application of these concepts to augment the capacities of graph databases, the next subsection presents the idea behind different types of graph databases.

2.4 Graph Databases

A graph represents entities as nodes and the relationships between them as edges. This enables us to model scenarios (Robinson et al. 2013) from various fields (e.g., sciences, semantic webs and social networks) and this allows for a natural form of representation (Jouilli and Vansteenberghe 2013; Miller 2013). The most popular graph model is the property graph (i.e., nodes contain properties, and relationships have names, are directed and can also contain properties) (Robinson et al. 2013).

Graph database management systems (referred to also as graph databases) take advantage of convenient graph features to manage (i.e., read, delete, update and even create) data and represent it as graph data models. Graph databases are very useful for data that are highly connected, as, unlike other database management systems, they can model relationships (Robinson et al. 2013) They are able to tackle the increasing complexity of data, which is one of the key challenges of big data. Thus, graph databases provide a viable alternative to relational databases (Jouilli and Vansteenberghe 2013; Miller 2013). Data that are structured as graphs makes it possible to query and analyze relationships and to model their attributes (Poulovassilis 2013).

There are three major factors that motivate us to use graph databases instead of other data storage systems: performance, flexibility, and agility. When dealing with connected and complex data, graph databases exhibit a much higher performance than relational databases and non-structured query language (NoSQL) stores. As graphs are additive, new relationships, nodes and subgraphs can be added easily (i.e., flexibility). Graph databases make it possible to develop data models parallel to other applications and software and thus to changing business environments (i.e., in an iterative process using incremental steps, i.e., agility Robinson et al. 2013). These factors show the close relationship of graph databases with the human brain (i.e., biological cognition).

The market space for graph databases is heterogeneous with vendors offering graph databases for different use cases. For example, AllegroGraphDB,[8] Hyper-GrahpDB,[9] OrientDB,[10] and Neo4j[11] are different graph databases. These are just a selection of databases available. However, the most popular one is Neo4j. In the next subsection, the capabilities of Neo4j are discussed in detail including an overview of the advantages and disadvantages.

2.4.1   Neo4j

Neo4j was introduced in 2003 and is one of the first NoSQL graph databases (Neo4j 2014a). Due to its Java fundamentals, it offers a versatile basis for creating a graph database. It is very flexible and useful for tagging, metadata annotations and social networks (Hoff 2009). Some of its main advantages are that it allows for atomicity, consistency, isolation, durability (ACID) transactions. It is highly available, able to handle billions of nodes and relationships; allows for very fast querying; and has its own graph query language, Cypher (Neo4j 2015a). Additionally, Neo4j applies both native graph processing and storage, and it supports PGM as a data model (Robinson et al. 2013). Neo4j is highly available when distributed among different servers, and it can be scaled easily when the amount of data is growing. Neo4j's query language is a declarative graph query language, which allows querying and updating in an efficient manner, and expresses complicated queries easily. Cypher is closely related to the Gremlin query language and supports the creation of graphs and the search for information. Many Cypher operators are inspired by SQL, which is a great advantage, and pattern matching operations are influenced by SPARQL protocol and RDF query language (SPARQL) terminology (Neo4j 2014b). Instead of only focusing on a simple writing structure, Cypher is reading-oriented. It applies an iconography, which is inspired by the way humans intuitively express graphs in diagrams (Robinson et al. 2013). Furthermore, complex search queries can be expressed and graph traversals are fast and efficient. In addition to Cypher, Neo4j supports traversals with RDF query language SPARQL and Gremlin. The system allows for schema-less structures, which provides increased flexibility when interventions during runtime are required. Finally, extensive documentation facilitates broad adaption.

---

[8]http://allegrograph.com/.
[9]http://www.hypergraphdb.org/index.
[10]http://orientdb.com/orientdb/.
[11]http://neo4j.com/.

Table 1 Overview of selected graph databases

Graph databases	Advantages	Disadvantages
AllegroGraph	Bridge gap between RDF and NoSQL	Insufficient PGM support
	Manual and documentation in good quality	Complex transfer object-based programming to RDF
	Easy partitioning	Blueprints interface is missing
		Not all OWL standards supported
HyperGraphDB	Java objects are stored easily	Terminology hard to understand
	Simultaneous connection of nodes, leads to more flexibility	Ad hoc queries pose problems
	Easy scalability within peer-to-peer network	Storage could be more efficient
	Efficient querying	Graph partitioning lack documentation
OrientDB	Simple installation	Community relatively small
	Multimode support	Out-dated manual and documentation
	Choice to work with or without schema	Problems when scaling large graphs with many nodes
	Supports local, distributed and embedded file systems	
Neo4j	High scalability	Performance depends on JVM
	Simple/powerful query language	No native multi-tenant support
	Intuitive data representation	
	Established solution	
	Large community	
	Create and search with Cypher	

One of the disadvantages of Neo4j is its dependency on Java because of the performance being heavily dependent on the quality of the java virtual machine (JVM) implementation. Furthermore, Neo4j does not have native multi-tenant support built in (Neo4j 2017).

2.4.2 Prototype with Neo4j

Although all of the above graph databases have similar properties and all of them have advantages for certain use cases, Neo4j is chosen as the basis for the proposed prototype as it provides particularly strong advantages.

First, it is currently the most successful, popular, and most used graph database. It is highly scalable, possesses a simple and powerful query language, and represents data intuitively. It supports both native graph storage and graph processing. Additionally, it is backed by a large community and is open-source for non-commercial use. Neo4j's versatility allows for its configuration as a server or as a Java embedded solution. It offers high performance and a vast array of interfaces to interact with different programming languages. Additionally, Neo4j is available through software as a service platforms (SAAS) (e.g., GrapheneDB[12]) where databases with up to 1,000 nodes and 10,000 relations are available for free. The query language Cypher is similar to SQL, it is relatively straightforward to learn, and its code is easy to interpret because its statements are based on iconography. The use of Cypher is not limited to Neo4j; the OpenCypher[13] initiative makes the access to the query language open source and enables graph processing with any product or application (Opencypher 2017). Furthermore, Neo4j provides fast and efficient query operations, scales well over different instances, and enables a simple visualization of graphs. Using Neo4j offers the possibility of exploring knowledge structures and finding related terms more efficiently (Portmann and Pedrycz 2014). Finally, Table 1 presents a short overview of the advantages and disadvantages of selected graph databases adapted from Edlich et al. (2011).

3 Prototype

This section presents the prototype and is divided in six Subsections. The first Sect. 3.1 explains the intention of the prototype through a scenario. Section 3.2 presents the intention of the prototype. The third Sect. 3.3 shows the interaction with the user and Sect. 3.4 presents the functionality of the prototype. The technical specifications of the prototype are shown in Sect. 3.5. Section 3.6 offers a synopsis of this section.

3.1 Scenario—A Tool for Cognitive Cities

This section introduces a fictional scenario to give an illustrative example for the prototype's implications as a tool and to explain its advantages.

John and his wife Megan spend their honeymoon travelling through Europe. John likes to organize their voyage by himself without pre-booking all tours in advance. The only bookings they made are for the hotels and the transportation between them. After arriving in Switzerland, they decide upon several sightseeing hotspots that they would like to visit. After a trip to Lucerne and to the Mountain of Jungfraujoch, they have about three hours remaining to visit Bern, with the "Altstadt" (old town) a UNESCO world heritage site.[14] John and Megan do not know Bern, and therefore they have no idea which places to visit in a given time frame.

---

[12]http://www.graphenedb.com.

[13]http://www.opencypher.org.

[14]http://whc.unesco.org/en/list/267.

The only suggestions they can find on the internet for sightseeing tours are static routes, where the starting and end points, the sightseeing spots and the duration of the tour are fixed and cannot be changed. However, John and Megan would prefer more flexibility and individuality during their tour. For example, they want the tour to start exactly at their hotel. Moreover, they cannot find a route that fits their time frame, as the tours are either too long or too short. They ask the hotel manager for advice and he shows them an innovative new website where they can design their own sightseeing route. To find a route that fits their preferences, John uses his laptop and accesses the tool. John enters their starting point (i.e., the "Berner Münster" as the hotel they are staying at is next to this sightseeing spot), the time frame they have at their disposal (i.e., "about three hours"), and their ending point (i.e., the "Rosengarten" as they meet there for dinner with a friend). The tool suggests several routes, satisfying their fuzzy criteria (i.e., "about three hours") and their fixed criteria (i.e., starting and ending point). Thus, the tool is able to provide solutions that fit the user's needs and preferences based on their fuzzy criteria. In the beta version of the tool, the benefits of the tools scalability will be applied and the tool will no longer only be applied for the city context but also for the country or the region context. Thanks to this web tool, John and Megan are able to make the most out of their visit in Bern and are happy with the visited sightseeing spots. The tool is a further step to establish Bern as a cognitive city.[15]

3.2   Route Planning Prototype

Citizens and travelers often rely on internet tools and mobile phone applications to plan their travel, e.g., Google Maps,[16] route planner on MySwitzerland,[17] OpenStreetMap[18] (Xiang et al. 2015). For example, the shortcomings of the routes suggested on MySwitzerland are that they are static and cannot be adapted to the users' needs. If a tourist has a specific time frame available, it would be useful if a tool could provide a route adapted to this time frame. Another useful feature would be if the starting point and end point were flexible. Finally, travelers might also want to visit specific sightseeing spots that are not included in the proposed, static routes. Sygic Travel is addressing some of these shortcomings and applying similar functionalities as the proposed prototype. One major differentiation between Sygic Travel and the prototype is the application of fuzzy logic and thereby the possibility to create routes based on vague terms (Chia et al. 2016).[19] Another differentiation is the possibility to select the time available during the days of the visit. This is possible in the prototype

---

[15]More information about the prototype and the scenario are available under https://smartandcognitivecities.blogspot.com.

[16]http://www.maps.google.com.

[17]http://www.myswitzerland.com/en-ch/home.html.

[18]https://www.openstreetmap.org.

[19]The applied logic of Sygic Travel is not publicly available.

but not in Sygic Travel. The prototype presented in this paper considers the duration of the route, its starting and end point, and the sightseeing spots as variables and is thus able to provide flexible routes. Instead of the classic Traveling Salesman Problem to visit every spot once in the shortest amount of time (Hoffman et al. 2013). The focus of the prototype lies in showing the functionality and the technical specifications. As argued in Sect. 2.4, the graph database Neo4j is used and augmented by the prototype. To translate numeric into linguistic values, (or to assign linguistic values to a range of numeric values) custom functions would be needed. However, Neo4j's query language Cypher does not support custom functions. To apply the proposed prototype, a standalone add-on for the graph database Neo4j is created, which can interact with Neo4j's backend and makes it possible to perform these calculations outside of Neo4j. The prototype is a Node.js application and a standalone add-on for Neo4j and extends its functionality. For this purpose, it uses various tools (i.e., node-neo4j application program interface [API[20]], Google Geocoding API, graph styling sheets) and Neo4j itself. The functionality is shown in the fourth Subsection followed by the technical specifications of the prototype.

3.3   Interaction with the User

The user of the prototype selects preferences (i.e., starting point, ending point) and then uses the query engine to submit a query. The prototype then processes the query. After the submission of the query the user receives an output that presents the given routes based on the preferences and the query. The user does not see the proceeding of the prototype. The figures in Sect. 3.5 show the visualization of the output shown to the user of the prototype.

3.4   Functionality of the Prototype

This subsection presents the functionality of the prototype. It shows how data can be read from and written to Neo4j, and how JSON map data can be read from third party API. Figure 1 visualizes the functionality of the prototype.

Step 1   The prototype receives a query.

Step 2   The prototype can read values provided in a custom API endpoint (e.g., google geocoding API) and acquire the required data to proceed. This happens before the execution of the files.

Step 3   The values have to be formatted in the javascript object notation (JSON) format to be readable by the prototype.

Step 4   The prototype can access Neo4j's backend through the node-neo4j module, which must be previously installed. Through this procedure, it can read values that are stored in the Neo4j database. This allows it to read either

---

[20]https://developers.google.com/maps/?hl=en.

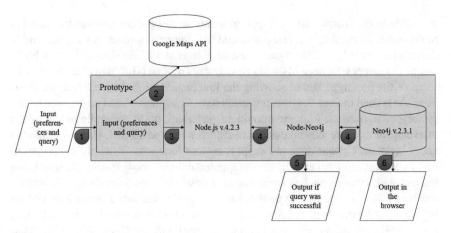

Fig. 1 Functionality of the prototype

complete nodes or specific information (e.g., some predefined values). The prototype can also write values back to the Neo4j database. This depends on the data that it reads before from a custom API endpoint or from the Neo4j database. The core functions of the prototype are reading data and generating content. More specifically, it executes three files, reads content (by reading nodes with certain IDs) or generates content

Step 5   After the execution of the three files a message in the console confirms the success or the failure of the operation. Thereby, the node-neo4j API is used to read nodes in the first action of the file execution. In the second action of the file execution, the prototype defines custom variables to transform the variables.

Step 6   In the last action of the file execution, the findings are reported back to the Neo4j database or new nodes are created using these transformed variables.

To understand how the prototype works, remember the user scenario from above: The prototype Neo4j database contains information about sightseeing spots (i.e., a certain number of nodes) which all contain an ID, a name, and a distance from a specific location (e.g., the train station.) The coordinates are derived from Google over the Google Geocoding API. The distances are expressed as numerical values. Currently, these sightseeing spots including their categorization (e.g., landmark, museum, park) are entered manually by the developer or administrator of the proto-type.[21]

The database also can be replicated into to a new database consisting of nodes, because the prototype has the ability to modify more than one instance.

The prototype has the advantage that it enhances knowledge representation, thus improving the user experience, as Neo4j has the ability to present nodes containing different information with different colors. Additionally, through the integration of

---

[21] Appendix A1 shows a list of all available landmarks and their grouping.

grass style sheets (which are identical to cascading style sheets (CSS)), the visual appearances of the nodes can be altered, for example by using different sizes. Additionally, the name of the node can be made to appear in a larger font through text formatting. These procedures all enhance the visual representation of nodes with labels.

One problem of using the distance in kilometers for the determination of the routes is that it does not contain all the relevant information. Consider two routes with an equal distance in kilometers, but one is characterized by many ascents. It will take more time to reach the endpoint of the route with the ascents. Therefore, the prototype uses duration (e.g., functionality "walking" in Google Maps) to determine the relation of the nodes. It is possible to select various methods of movement (e.g., walking, public transportation, car). The prototype currently runs on a desktop computer connected to a server. The technical functionality is presented in the next subsection.

3.5 Technical Functionality of the Prototype

The prototype consists of three sub-programs (i.e., Neocreate.js, Neorel.js, Neofind.js) to ensure the desired functionalities as seen in Fig. 1. This Subsection presents the three files in detail.

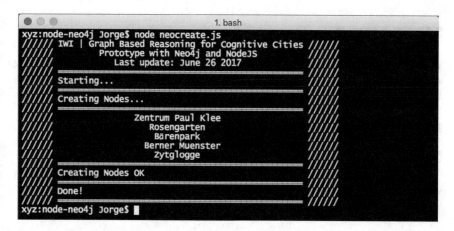

Fig. 2 Output for the input scenario

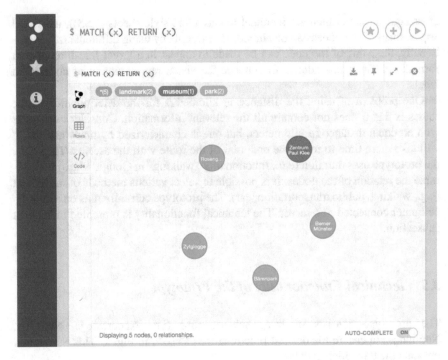

Fig. 3 Unconnected nodes

3.5.1 Neocreate.js

The functionality of the Neocreate.js file is to create the nodes. Therefore, the file is connected with the Neo4j database and creates the nodes. The nodes have two attributes: *name* and *GPS coordinates*.

Example 1: *CREATE (ZentrumPaulKlee:museum{name: "Zentrum Paul Klee", location:46.948741,7.474280})*.

After the execution of this file a short output is created to indicate if the execution was successful. Figure 2 shows an output for the input scenario.

The creation of the nodes is also visible in the browser (i.e., step 6). Figure 3 shows the created nodes through Neocreate.js.

3.5.2 Neorel.js

Example 2: var zenros = http://maps.googleapis.com/maps/api/directions/json?orig in=46.948741,7.474280&destination=46.951868,7.460365&sensor=false&mode= walking.

Then, the duration (i.e., in minutes) of the relation between the nodes is processed.

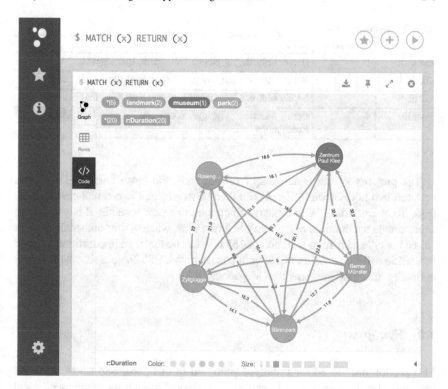

Fig. 4 Relationships between nodes

Example 3: *db.insertRelationship(idzen, idros, 'r:Duration', {duration: ' '+zenrosrtime+' '}.*

The Neo4j database now consists of n2-n relations with n = number of nodes. This inherent complexity is an example for the fit of a graph database (see Sect. 2.4) for this prototype. The distance between two nodes can vary (e.g., steep ascents or contorted alleys). The creation of the relationships is equally visible in the browser as seen in Fig. 4.

3.5.3 Neofind.js

The third file seeks the shortest path between the desired nodes. In this prototype, we apply the shortest path method available in Neo4j. Note: The Dijkstra-Algorithm (Stein et al. 2001) would be another possible method to determine the shortest path between nodes. In order to find the shortest path between a set of nodes, Neo4j requires the following prerequisites: A starting and ending node need to be set. Afterwards, a path between two nodes a and b is established, where path p = (a) − [r*] − (b) is a set of nodes connected by the relationship r (Neo4j 2015b).

Table 2 Proposed routes

Fit	Route no.	Nodes
Good fit	1	Berner Münster, Zytglogge, Bärenpark, Rosengarten
Ok fit	2	Berner Münster, Zytglogge, Zentrum Paul Klee, Rosengarten
Ok fit	3	Berner Münster, Bärenpark, Zentrum Paul Klee, Rosengarten
Poor fit	4	Berner Münster, Zytglogge, Bärenpark, Zentrum Paul Klee, Rosengarten

The property of the relationship r (in our case, the travel duration in minutes between two nodes seen in Fig. 4) can then be aggregated to create a new variable (e.g., Total Duration.) With a query which uses this new variable, it is possible to approximate the duration of a journey to a fuzzy set, where a time constraint such as "three hours" and a Total Duration of 185 min can be matched to constitute a "good fit", while a Total Duration of 195 min equals an "ok fit". Table 2 shows the routes optimally matching the criteria of Megan and John.

3.6 Synopsis

This section introduced the prototype proposed in this paper and suggested a use case scenario with Megan and John visiting Bern. Through the functionality and the benefits of the presented prototype Megan and John had an increased level of flexibility on their trip. The possibility to create dynamic routes for sightseeing trips offers new possibilities to the travel industry. The possibility to add specific time exposures to the sightseeing spots improves the provided routes. Thereby, visitors can choose between actively discovering a sightseeing spot or stopping only for a short time. In connection with cognitive cities (i.e., the city of Bern is starting to become a smart city as a preface of a cognitive city) this enhanced interaction between all stakeholders of a cognitive city (e.g., visitors, tourism office) is a win-win situation for all. The next Section presents the conclusion and the outlook of the paper.

4 Conclusion

The ever-increasing urban population calls for efficient solutions to ensure cooperative living and a sustainable standard of living. This can be made possible by transforming cities into cognitive cities. To pursue this goal, a well-functioning interaction between humans and computer systems is crucial. As computer systems cannot compute imprecise and uncertain information (i.e., described in natural language), fuzzy logic is needed. The prototype proposed in this paper uses fuzzy logic to extend graph databases and provides users with a dynamic route planning tool. This allows them to account for fuzziness (i.e., uncertainties and imprecisions.), which enables graph databases to process even more information and queries and makes enhanced interaction between humans and computer systems possible.

For now, the rules that determine which linguistic term is assigned to which numeric values are defined manually. Future tools could use fuzzy if-then-rules (Neo4j 2015b) that are used by humans to transform numbers into linguistic data. Applying these rules to computer systems can make it possible to automatically determine the rules on how numeric values are translated into linguistic values.

This paper presents a prototype that is developed to apply the proposed idea to the travel industry. It offers a standalone process to interact with the graph database Neo4j. The prototype and thus the graph database are able to handle fuzziness. This makes it possible to create routes that are tailored to the needs of the traveler (i.e., based on his fuzzy preferences) considering starting and end point, and time frame he has available. This is an advantage over current tools and websites that only propose static routes. This shows how fuzzy logic can be used to develop tools that are able to cope with the needs and wishes of people and find solutions that are adequately tailored to them. The proposed idea can be applied not only to the travel industry, but also other areas of application could be possible (e.g., communication, work.) By streamlining the interaction between humans and computer systems, these dimensions can be enhanced considerably. More efficient, sustainable and resilient systems can be created, thus ultimately leading to cognitive cities (Portmann and Finger 2016).

The next steps include the building of a fully functioning tool based on the prototype, for example a mobile phone application. Furthermore, the prototype has to be applied not only to the specific case of the city of Bern in Switzerland but also to other cities (or metropolitan areas) in the world. This scalability as well as the flexibility of the interaction with the Google Geocoding API are major benefits of the prototype. The improvement of the functionalities (e.g., omitting manual processes of the prototype) as well as the user experience (e.g., improvement of the graphical user interface (GUI)) are top priorities for future research. Additionally, further research is needed to expand the proposed prototype to be useful not only for the travel industry but also for other aspect of daily life. For these purposes, the requirements of the users have to be assessed, to ensure user-friendly tools. Only by building well-functioning and intuitive tools can the goal of developing cities into cognitive cities be attained. The evaluation of the currently applied shortest path method function of Neo4j against other possible applications (e.g., the Dijkstra-Algorithm Stein et al. 2001) would be one measure to enhance the prototype. Another future research topic would be to generate various access points for the prototype (e.g., mobile phones, tablets) and to evaluate them.

Acknowledgements We would like to thank the student assistants of the Institute of Information Systems of the University of Bern for their valuable input and their contributions to this paper.

Appendix A. Additional Information of the Prototype

This Appendix A shows various additional information to the applied prototype.

A.1. Sightseeing Spots in Bern

See Tables 3 and 4.

Table 3 Sightseeing spots in Bern

Name	Name in the prototype	Group	Time exposure (min)	Location	Description
Zentrum Paul Klee	ZentrumPaulKlee	Museum	60	'46.948741, 7.474280'	A center dedicated to the painter Paul Klee[a]
Rosengarten	Rosengarten	Park	30	'46.951868, 7.460365'	A park with a beautiful view of the old town of Bern
Bärenpark	Baerenpark	Park	30	'46.947605, 7.459363'	A park where bears live[b]
Berner Münster	BernerMuenster	Landmark	20	'46.95098, 7.47318'	A cathedral in the old town
Zytglogge	Zytglogge	Landmark	20	'46.947973, 7.447796'	A medieval tower in the old town

[a]http://www.zpk.org/en/index.html
[b]http://www.baerenpark-bern.ch/index.php?id=info&L=0

Table 4 Route fit criteria

Fit	Range
Poor fit	10% or more variance
Ok fit	Less than 10% variance
Good fit	Less than 5% variance

A.2. Query Processing

This appendix shows how the prototype processes queries for a better understanding of the prototype. Every time the prototype generates nodes, these nodes obtain individual IDs. These IDs vary from node generation to node generation. Therefore, the prototype has to be able to identify the IDs of the nodes to recognize the underlying

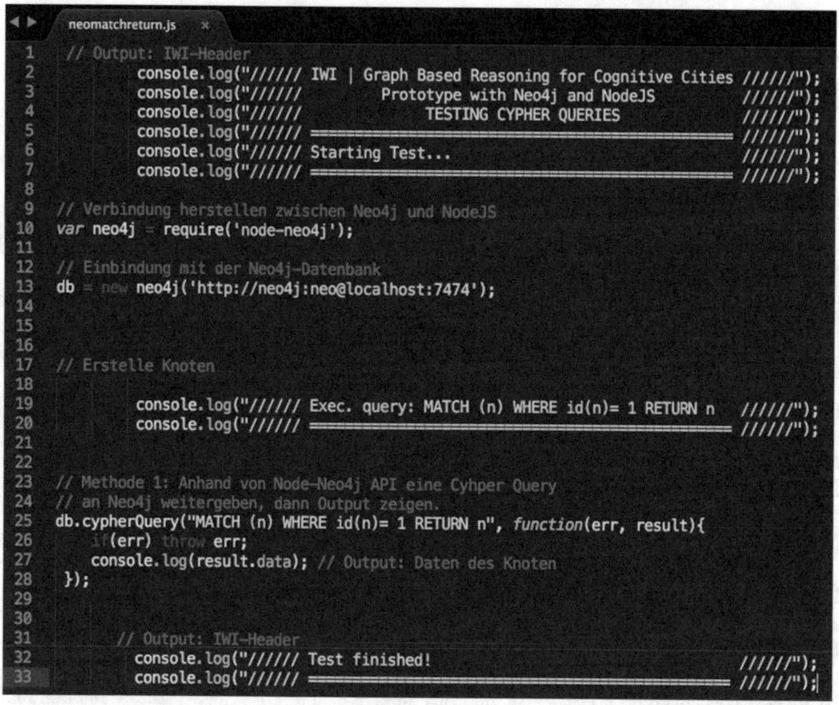

```
neomatchreturn.js   ×
1   // Output: IWI-Header
2       console.log("////// IWI | Graph Based Reasoning for Cognitive Cities //////");
3       console.log("//////           Prototype with Neo4j and NodeJS          //////");
4       console.log("//////                TESTING CYPHER QUERIES               //////");
5       console.log("////// ==================================================== //////");
6       console.log("////// Starting Test...                                     //////");
7       console.log("////// ==================================================== //////");
8
9   // Verbindung herstellen zwischen Neo4j und NodeJS
10  var neo4j = require('node-neo4j');
11
12  // Einbindung mit der Neo4j-Datenbank
13  db = new neo4j('http://neo4j:neo@localhost:7474');
14
15
16
17  // Erstelle Knoten
18
19      console.log("////// Exec. query: MATCH (n) WHERE id(n)= 1 RETURN n  //////");
20      console.log("////// ==================================================== //////");
21
22
23  // Methode 1: Anhand von Node-Neo4j API eine Cyhper Query
24  // an Neo4j weitergeben, dann Output zeigen.
25  db.cypherQuery("MATCH (n) WHERE id(n)= 1 RETURN n", function(err, result){
26      if(err) throw err;
27      console.log(result.data); // Output: Daten des Knoten
28  });
29
30
31      // Output: IWI-Header
32      console.log("////// Test finished!                                       //////");
33      console.log("////// ==================================================== //////");
```

Fig. 5 Neo4j ID identification

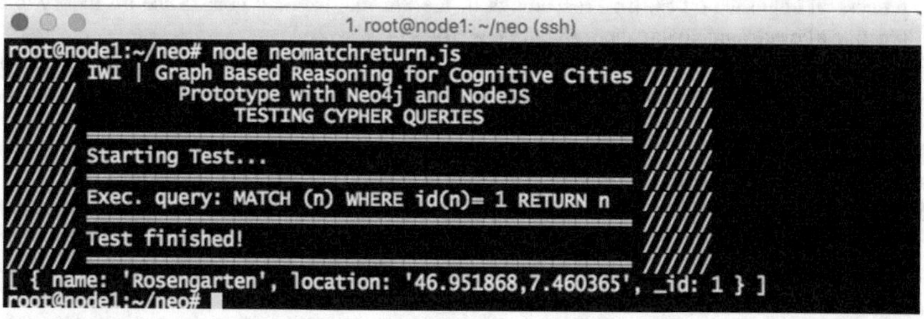

```
●  ●  ●                    1. root@node1: ~/neo (ssh)
root@node1:~/neo# node neomatchreturn.js
//////  IWI | Graph Based Reasoning for Cognitive Cities //////
//////           Prototype with Neo4j and NodeJS          //////
//////                TESTING CYPHER QUERIES               //////
////// ====================================================
////// Starting Test...                                    //////
////// ====================================================
////// Exec. query: MATCH (n) WHERE id(n)= 1 RETURN n      //////
////// ====================================================
////// Test finished!                                      //////
////// ====================================================
[ { name: 'Rosengarten', location: '46.951868,7.460365', _id: 1 } ]
root@node1:~/neo#
```

Fig. 6 Neo4j console

attributes of the nodes. Figure 3 in Sect. 3.5.1 shows a graph with nodes and Fig. 5 shows the identification of the nodes.

As seen in row 10 of Fig. 5 Neo4j creates a connection with Node.js and node-neo4j API. Row 13 shows the inclusion of the Neo4j database and row 25 shows the Cypher query which will be passed on to Neo4j "MATCH (n) WHERE id(n) = 1 RETURN n", which means "show me all nodes with the ID n where n = 1." Row 27 shows the redirection of the output to the console as seen in Fig. 6. This output is visible to the user.

References

Boisson de Marca JR (2015) Smarter Cities... and Wiser Ones? In: IEEE The Institute, vol 38

Caragliu A, Del Bo C, Nijkamp P (2011) Smart cities in Europe. J Urban Technol 18(2):65–82

Chia WC, Yeong LS, Lee FJX, Ch'ng SI (2016) Trip planning route optimization with operating hour and duration of stay constraints. In: 11th international conference on computer science and education (ICCSE). IEEE, pp 395–400

D'Onofrio S, Zurlinden N, Portmann E, Kaltenrieder P, Myrach T (2016) Synchronizing mind maps with fuzzy cognitive maps for decision-finding in cognitive cities. In: Proceedings of the 9th international conference on theory and practice of electronic governance. ACM, New York, pp 363–364

Edlich S, Friedland A, Hampe J, Brauer B, Brückner M (2011) NoSQL: Einstieg in die Welt nichtrelationaler Web 2.0 Datenbanken. Carl Hanser Verlag, München

Goldstein J (1999) Emergence as a construct: history and issues. Emergence 1:49–72

Hoff T (2009) Neo4j—a graph database that kicks buttox. http://highscalability.com/neo4j-graph-database-kicks-buttox. Accessed 10 June 2014

Hoffman KL, Padberg M, Rinaldi G (2013) Traveling salesman problem. Encyclopedia of operations research and management science. Springer, New York, pp 1573–1578

Hurwitz JS, Kaufman M, Bowles A (2015) Cognitive computing and big data analytics. Wiley, Indianapolis

IBM (2015) The four V's of big data. http://www.ibmbigdatahub.com/sites/default/files/infograph ic_file/4-Vs-of-big-data.jpg. Accessed 21 May 2015

Jouilli S, Vansteenberghe V (2013) An empirical comparison of graph databases. In: International conference on social computing (SovialCom). IEEE, pp 708–715

Kaltenrieder P, Portmann E, D'Onofrio S, Finger M (2014) Applying the fuzzy analytical hierarchy process in cognitive cities. In: Proceedings of the 8th international conference on theory and practice of electronic governance. ACM, New York, pp 259–262

Kaltenrieder P, Portmann E, D'Onofrio S (2015a) Enhancing multidirectional communication for cognitive cities. In: Second international conference on eDemocracy and eGovernment. IEEE, pp 38–43

Kaltenrieder P, Portmann E, Myrach T (2015b) Fuzzy knowledge representation in cognitive cities. In: 2015 IEEE international conference on fuzzy systems (FUZZ-IEEE). IEEE, pp 1–8

Kaltenrieder P, Papageorgiou E, Portmann E (2016) Digital personal assistant for cognitive cities: a paper prototype. In: Portmann E, Finger M (eds) Towards cognitive cities: advances in cognitive computing and its applications to the governance of large urban systems. Springer, Berlin, pp 101–121

Kaufmann MA, Portmann E (2015) Biomimetics in design-oriented information systems research. In: At the vanguard of design science: first impressions and early findings from ongoing research,

research-in-progress papers and poster presentations from the 10th international conference, Dublin, pp 53–60

Kaufmann MA, Portmann E, Fathi M (2012) A concept of semantics extraction from web data by induction of fuzzy ontologies. In: International workshop on uncertainty reasoning for the semantic web. IEEE, pp 1–6

Kelly JE, Hamm S (2013) Smart machines IBM's Watson and the Era of cognitive computing. Columbia University Press, New York

Lin TY (1997) Granular computing: from rough sets and neighborhood systems to information granulation and computing in words. In: European congress on intelligent techniques and soft computing, pp 1602–1606

Miller J (2013) Graph database applications and concepts with Neo4j. In: Proceedings of the southern association for information systems conference, vol 2324, pp 141–147

Mostashari A, Arnold F, Mansouri M, Finger M (2011) Cognitive cities and intelligent urban governance. Netw Ind Q 13:3–6

Neo4j Manual (2014b) v2.1.2. http://docs.neo4j.org/chunked/milestone/cypher-introduction.html. Accessed 25 March 2014

Neo4j Manual (2015a) v2.2.3. http://neo4j.com/docs/2.2.3/introduction-highlights.html. Accessed 25 June 2015

Neo4j Manual (2017) v3.2. https://neo4j.com/docs/operations-manual/3.2/security/authentication-authorization/native-user-role-management/native-roles. Accessed 26 June 2017

Neo4j Website (2014a) http://www.neo4j.org. Accessed 7 June 2014

Neo4j Website (2015b) http://console.neo4j.org/?id=811v89. Accessed 31 Dec 2015

Opencypher Website (2017) http://www.opencypher.org. Accessed 26 June 2017

Perticone V, Tabacchi ME (2016) Towards the improvement of citizen communication through computational intelligence. In: Portmann E, Finger M (eds) Towards cognitive cities: advances in cognitive computing and its applications to the governance of large urban systems. Springer, Berlin, pp 83–100

Portmann E, Finger M (eds) (2016) Towards cognitive cities: advances in cognitive computing and its applications to the governance of large urban systems. Springer, Berlin

Portmann E, Pedrycz W (2014) Fuzzy web knowledge aggregation, representation, and reasoning for online privacy and reputation management. In: Papageorgiou E (ed) Fuzzy cognitive maps for applied sciences and engineering. Springer, Berlin, pp 89–105

Portmann E, Kaufmann MA, Graf C (2012) A distributed, semiotic-inductive, and human-oriented approach to web-scale knowledge retrieval. In: Proceedings of the 2012 international workshop on web-scale knowledge representation, retrieval and reasoning. ACM, New York, pp 1–8

Poulovassilis A (2013) Database research challenges and opportunities of big graph data. In: Gottlob G, Grasso G, Olteanu D, Schallhart C (eds) Big data. Springer, Berlin, pp 29–32

Robinson I, Webber J, Eifrem E (2013) Graph databases. O'Reilly Media, Sebastopol

Siemens G (2005) Connectivism: a learning theory for the digital age. Int J Instr Technol Distance Learn 2:3–10

Stein C, Cormen T, Rivest R, Leiserson C (2001) Introduction to algorithms, vol 3. MIT Press, Cambridge

United Nations Department of Urban and Social Affairs (2008) World urbanization prospects: The 2007 revision—executive summary. http://www.un.org/esa/population/publications/wup2007/2007WUP_ExecSum_web.pdf. Accessed 25 June 2015

Xiang Z, Magnini VP, Fesenmaier DR (2015) Information technology and consumer behavior in travel and tourism: insights from travel planning using the internet. J Retail Consum Serv 22:244–249

Yao YY (2000) Granular computing: basic issues and possible solutions. In: Proceedings of the 5th joint conference on information sciences, vol 1, pp 186–189

Yao YY (2005) Perspectives of granular computing. In: International conference on granular computing, Beijing, pp 85–90

Zadeh LA (1965) Fuzzy sets. Inf Control 8:338–353

Zadeh LA (1979) Fuzzy sets and information granulation, advances. In: Gupta M, Ragade RK, Yager RR (eds) Fuzzy set theory and applications. North-Holland Publishing, Amsterdam, pp 3–18

Zadeh LA (1988) Fuzzy logic. Computer 21:83–93

Zadeh LA (1997) Toward a theory of fuzzy information granulation and its centrality in human reasoning and fuzzy logic. Fuzzy Sets Syst 90:111–127

Zadeh LA (1999) From computing with numbers to computing with words—from manipulation of measurements to manipulation of perceptions. IEEE Trans Circuits Syst I Fundam Theory Appl **45**, 105–119

Zadeh LA (2004) Precisiated natural language (PNL). AI Mag 25:74–91

Patrick Kaltenrieder is a course-manager, lecturer and senior consultant for the digital transformation, online marketing and business modeling. At present, he works as senior consultant at the Internetagentur iqual and as course-manager and lecturer at the University of Bern. After his Business Administration Bachelor's and Master's Degree at the University of Bern, Patrick Kaltenrieder worked for about two years at a major Swiss Bank and then returned to the University of Bern for his Dissertation. He obtained his Ph.D. in Business Administration, focusing on smart and cognitive cities, on knowledge aggregation, representation and reasoning, and on stakeholder management and cognitive computing. During his Ph.D.-studies and his active research phase Patrick Kaltenrieder has written a blog about his smart and cognitive cities research to keep the readers of his papers in the loop. The blog is still available at http://smartandcognitivecities.b logspot.ch/. Patrick Kaltenrieder is the author of several papers published at various International Conferences such as the FUZZ-IEEE 2015 in Istanbul and the FUZZ-IEEE 2016 in Vancouver.

Jorge Parra is an IT and management consultant based in Bern. He studied for a BS.c. in Business Administration at the University of Bern and for an MS.c. in Business Administration and Information Systems at the University of Bern and at the National Taiwan University. During his master's studies, he was awarded the first prize at the Accenture Campus Challenge 2013 for developing an Indoor Mobile Strategy for airports. Under the supervision of Edy Portmann, he wrote his master's thesis about knowledge reasoning and representation with graph databases and explored the use of the graph database Neo4j with fuzzy logic in conjunction with Node.js, a JavaScript environment for executing code server-side. Furthermore, he assisted Edy Portmann with his research on cognitive cities and soft computing. He has a keen interest for IoT and Blockchain technology as well as cloud computing. Nowadays he is helping large enterprises transit into the era of digitalization, transforming their organizations and managing their software development process to be more efficient applying agile methodologies. He is currently working at APP Unternehmensberatung AG, a management consulting company, is passionate about travelling, soccer and mountaineering and lives in Bern.

Thomas Krebs is working as an enterprise architecture specialist and a project manager for a start-up in Solothurn, Switzerland. Currently, he advises a client on the introduction of a competence-based Job-Matching System. Thomas Krebs uses practical and methodological expertise in business analysis and requirements engineering to help the clients achieve their business goals. His professional career started in an international consulting firm, advising large corporations on outsourcing and IT-transformation projects. He worked for clients in the telecommunications and textile industry, applying expert knowledge in IT-Service Management. Various assignments required frequent travel abroad to ensure the success of the projects. In the past, Thomas Krebs studied for a BSc in Business Administration and a MSc in Business IT at the University of Bern, Switzerland. During his studies he spent two semesters abroad for international exchanges at the Turku School of Economics, Finland, and the Université Panthéon-Assas in Paris, France. In his Master's Thesis, he analyzed fuzzy cognitive maps in the context of the semantic web. During the work for his thesis, he acquired in-depth knowledge of graph databases and how to apply them to fuzzy problems. In his spare time, he is a passionate musician and loves the outdoors. Thomas Krebs is happily married and lives near Zurich.

Noémie Zurlinden is a PhD student in Economics at the University of St. Gallen, Switzerland. Her research interests lie in the fields of development, cultural and historical economics. She completed the Swiss Program for Beginning Doctoral Students in Economics at the Study Center Gerzensee. Before starting her Ph.D., she got a Bachelor's degree in Economics, Social Sciences and Philosophy, and a Master's degree in Economics, both from the University of Bern, Switzerland. During her Master's, she worked part-time as a research assistant at the Institute of Information Systems, University of Bern. She also worked at the Busara Center for Behavioral Economics in Nairobi, Kenya, as a research assistant for three months.

Edy Portmann is a researcher and scholar, specialist and consultant for semantic search, social media, and soft computing. Currently, he works as a Swiss Post-Funded Professor of Computer Science at the Human-IST Institute of the University of Fribourg, Switzerland. Edy Portmann studied for a BS.c. in Information Systems at the Lucerne University of Applied Sciences and Arts, for an MS.c. in Business and Economics at the University of Basel, and for a Ph.D. in Computer Sciences at the University of Fribourg. He was a Visiting Research Scholar at National University of Singapore (NUS), Postdoctoral Researcher at University of California at Berkeley, USA, and Assistant Professor at the University of Bern. Next to his studies, Edy Portmann worked several years in a number of organizations in study-related disciplines. Among others, he worked as Supervisor at Link Market Research Institute, as Contract Manager for Swisscom Mobile, as Business Analyst for PwC, as IT Auditor at Ernst and Young and, in addition to his doctoral studies, as Researcher at the Lucerne University of Applied Science and Arts. Edy Portmann is repeated nominee for Marquis Who's Who, selected member of the Heidelberg Laureate Forum, co-founder of Mediamatics, and co-editor of the Springer Series 'Fuzzy Management Methods', as well as author of several popular books in his field. He lives happily married in Bern and has three lively kids.

Thomas Myrach is Full Professor for Business Administration at the University of Bern, Switzerland (since 2002). He is Director of the Institute of Information Systems and leads the group Information Management. He has studied Business Administration and Computer Science at the University of Kiel, Germany. His Doctoral degree and the Venia legendi he obtained from the University of Bern, Switzerland, with theses on Data Dictionaries and Temporal Databases. In the years 2001/2002 he has been visiting Professor for e-business at the Technical University of Aachen, Germany. In research and teaching he is concerned with the potentials and challenges of digitalization in business. In particular, he has worked on the vision of E-Business and the changes induced by network technologies like the internet. More recent interests include topics like digital sustainability, open data and ICT procurement. He has a keen interest in interdisciplinary and transdisciplinary work. As Vice-President of the Collegium Generale at the University of Bern he has been responsible for several transdisciplinary lectures. He is author and co-editor of several books and publishes regularly in international journals.

Correction to: Possibilities for Linguistic Summaries in Cognitive Cities

Miroslav Hudec

Correction to:
Chapter "Possibilities for Linguistic Summaries in Cognitive Cities" in: E. Portmann et al. (eds.), *Designing Cognitive Cities*, Studies in Systems, Decision and Control 176, https://doi.org/10.1007/978-3-030-00317-3_3

The affiliation "Faculty of Economic Informatics, University of Economics in Bratislava" of author "Prof. Miroslav Hudec" in the original version of the book has been changed to "Faculty of Organizational Sciences, University of Belgrade" in the chapter "Possibilities for Linguistic Summaries in Cognitive Cities". The correction book has been updated with the change.

The updated version of this chapter can be found at
https://doi.org/10.1007/978-3-030-00317-3_3

Glossary

Action Design Research Action Design Research (ADR) is a concept in the information sciences that tries to bring research and practice into the best possible exchange. It combines design science with action research. In this view IT-artefacts are 'ensemble artefacts' originating from the interplay of design and the context in which design takes place which is influenced by the values and assumptions of a wide variety of communities of developers, investors and users. Building and evaluating ensemble IT-artefacts culminate in prescriptive design knowledge.

   Citizen Communication The term 'citizen communication' usually refers to all human-to-human and human-to-group communication whose aim is to deal with collective sharing of opinions, feelings and resources in metropolitan and rural areas. Improvement of Citizen Communication through computational intelligence can increase citizen participation and contribute in the mitigation of a general sense of apathy that has pervaded social life in the recent history of civilization.

   Cognition Cognition is the faculty of the human mind to represent the world on the basis of sensual perception, to categorize it and to communicate its abstraction to others. Saying that "the child has cognitive deficiencies" usually means that his or her perception and interpretation of the word is anomalous. The semantic field of the term cognition contains mainly two approaches: cognition as a result and cognition as a process. As a result, cognition is considered as the end of the path of making the world intelligible. This is equivalent to knowledge. As a process, cognition is the road to walk for accessing the knowledge of the world.

   Cognitive Cities A cognitive city is an enhancement of a (human) smart city. The concept, however, does not replace smart city-approaches but complements them by focusing on a specific aspect of the smart city: interaction and communication between the stakeholders and the city. Cognitive cities rely on well-connected networks of urban systems that senses through collecting data, learn in real time, and communicate their learnings with all city stakeholders. The technical basis are cognitive computer systems, which are capable of recognizing patterns in huge amounts of data and learn by interacting and communicating with the people who use them. Developments such as the Internet of Things (IoT), cloud-based social feedback, crowdsourcing, and predictive analytics allow cities to actively and independently

E. Portmann et al. (eds.), *Designing Cognitive Cities*, Studies in Systems, Decision and Control 176, https://doi.org/10.1007/978-3-030-00317-3

learn, build, search, and expand when new information is added to the already existing ones. Thus, the cognitive city is able to develop collective and humanistic intelligence. The concept aims to answer demands of future cities that cannot be met by the means of efficiency and sustainability only, but also address resilience as well as the citizens need for participation and individualism.

Cognitive Computing Cognitive computing is enhancing computing with cognition. Cognitive computing has various subjects encompassing computational thinking, connectivism, as well as the intelligence amplification loop. Cognitive computing systems evolve and are able to learn based on data and experience through the detection of patterns and thereby deriving meaningful information from multiple sources (e.g., images, texts).

Cognitive Maps Resemblance Resemblance is the similarity in meaning of two or more terms. Linguistic relations of synonymy, antonymy, hyponymy or hypernymy show similar or dissimilar meaning between pairs of words and similarity measures provide tools to calculate the neighborhood or distance between them. Names of social relations may be similar in meaning, have opposite meaning or to participate in meaning inclusion. Centrality measures, offering a calculation of the relevance of a node in a network, and similarity measures, quantifying the resemblance of words, permit to calculate how one cognitive map (a visual representation of a problem with causal dependencies) is similar to another.

Collaborative Intelligence Collaborative Intelligence is a new paradigm of Artificial Intelligence whereas the interaction between people and machine is lifted to a paritary level to solve complex problems. In CI, computers partner with people to achieve the common goals. The assumption is that some subtasks are more reasonably delegated to the person, and others to the computer. Thereby, a CI system is intended not to substitute people but to collaborate with people.

Collective Intelligence—Urban Intelligence Collective intelligence consists of individual intelligences: more or less intelligent human beings as well as more or less intelligent objects such as Artificial Intelligences (AI) or electronic devices. They form a network that is more than the sum of its elements and contributes to problem solving differently than individual intelligences would. Urban intelligence denotes the emergence of collective intelligence within a city. Therefore, incoming (novel) civic data enables to transform municipalities' formerly sparse knowledge to a much more sophisticated understanding of cities. This intelligence creates unique and valuable insights from the city to foster its development.

Computational Intelligence (CI) is an umbrella term coined in the 00's to regroup all the methodologies and techniques that try to solve problems in contexts of ambiguous, incomplete, missing or vague information using approaches that are often derived from 'natural' methods, such as the ones devised by human minds or evolved in nature from animal behavior. CI methods are generally aimed at suboptimal solutions that can nonetheless be achieved in a reasonable timeframe, and that are 'good enough' for the intended problem. Some of them are directly inspired by human reasoning, especially among the earliest instances (such as Fuzzy Logic, Soft Computing and the likes).

Computational Theory of Perceptions (CTP) is an attempt to formalize in a way that is computationally feasible the capability of humans to perform a wide variety of physical and mental tasks without any measurements and any computations. CTP enables the representation and processing of perceptions, by using Fuzzy Set Theory as the mathematical underpinning to deal with intrinsic graduality and granularity of perception-based information.

Computing with Words (CWW) aims to replace the intermediate step of symbolic logic to offer a perspective of computation that is more directly linked to the way humans do their internal reasoning. CWW offers a fusion between natural language, and specifically its verbal characteristics, and computation by standard fuzzy variables. Basic information manipulated in CWW is a collection of propositions expressed directly in words from a natural language, as a human would do. This is in contrast with classical computation, which is based on a three-steps model of transforming input into numbers or logic predicates, operating on such abstract objects and then transforming back the output of computation on something meaningful for humans.

Conceptual Energy Model The Conceptual Energy Model describes the urban energy system as three sub-layers: the human-institutional layer, the physical layer, and the data layers. The Model provides elaborated mechanisms for energy systems efficiency in a cognitive city.

Conjecture In mathematics, a conjecture is a conclusion, or a proposition based on incomplete information, for which no proof or sufficient evidence has been found. Conjectures are classified into consequences, hypotheses and speculations.

Connectivism Connectivism denotes a recent cognition and learning theory that acknowledges technology's impact on human learning. It indicates that it is no longer sufficient for humans to exclusively learn by means of their own perceptions and experiences, but that they need to consult resources of other data bases, people and organization.

Creativity (Creative Reasoning) Creativity is considered at the top of human intellectual activity. It is related to what in ordinary human reasoning can be called "creative jumps" and "dangerous shortcuts". Creativity arises through a type of reasoning opposed to formal deduction, which is supposed to be performed without jumps and shortcuts.

Dimensional Data Structure Dimensional data structure is a data warehouse structure, where a central table represents the fact on which the analysis is focused, and a number of tables represent the dimensions required for analyses.

Embodied Cognition Embodied cognition is a view of cognition that stresses the role of the body in cognitive activities, such as perception or abstraction. Embodied cognition is indebted to the embodiment thesis, according to which many characteristics of cognition depend on the body (including the brain but not limited to it) conferring to it a constitutive or a causal role in the generation of knowledge, as the body may act as a restrictor, distributor or regulator of the cognitive activity. If a constitutive role is assigned, cognition is nothing but body cognition; if causal, the body simply helps to explain cognition.

Energy Domain Energy is one of city domains where efficiency, sustainability must be addressed. The aim is to improve energy resources management and to integrate buildings and infrastructures in the future Smart Grid. The Smart Grid is envisioned as the future generation grid, as it will integrate Information and Communication Technologies (ICT) to the existing power grid. This integration will enable a scalable incorporation of Distributed Energy Sources (DERs) to the current grid two-way communication network between energy producers and consumers. The realization of the Smart grid approach requires learning and cognitive cities to improve current grid efficiency, sustainability and resilience.

Energy Resilience Energy resilience consists on the use of cognitive systems in collaboration with humans in order to prevent, avoid and react to power outages caused by power peak periods or natural disasters. Energy Demand Response (DR) applications and power outage management systems (OMSs) are two of the best known solutions required to improve the energy resilience in the future Smart Grid.

Energy Sustainability Energy sustainability consists on the improvement of energy management by a) providing citizens (i.e., energy consumers, energy auditors, building designers) a complete assessment of city infrastructure energy performance and b) suggesting citizens actions to change their energy management behavioural patterns for economic, social and ecological purposes through learning systems.

Fashion A popular or the latest style of clothing, hair, decoration, or behaviour.

Folksonomy A folksonomy or social tagging is a collection of user-defined metadata that evolves through time and while users create or store content to identify what they think the content is about; "Tag clouds" pinpoint various identifiers and the frequency of use on a folksonomy.

Fuzzy Data is a non-precise number or category. A fuzzy number is a generalization of a real number by a convex fuzzy set, where each possible value has its membership degree between 0 and 1. A fuzzy categorical variable can take one of a fixed number of possible values as well as intensities of these values.

Fuzzy Logic Fuzzy Logic (in the wide sense) is a paradigm where concepts are admitted applying up to a degree. It is a form of many-valued logic in which the truth values of predicates may be any real number between 0 and 1. While classical sets are defined by 'sharp' boundaries (i.e., being a member or not), fuzzy sets enable having a gradual membership with 'soft' boundaries. Linguistic variables can be used as labels for the membership degree (e.g., weak, high). Compared to traditional methods of calculation, which strive for precision and absolute truth, fuzzy logic thus allows to include structured, semi-structured and non-structured data elements to handle natural language. Fuzzy Logic attempts to model perceptions in a computational form in order to give computers a human-like intelligence. The concept facilitates common-sense reasoning with imprecise predicates expressed as fuzzy sets.

Fuzzy Quantifier A Fuzzy Quantifier allow us to express imprecise quantities or proportions in order to provide an approximate idea of the number of elements of a subset fulfilling a certain condition. These quantifiers can be absolute and relative.

Fuzzy Set A Fuzzy Set is an extension of classical set. In classical set theory, the membership of elements to a set is assessed in binary terms; an element either belongs or does not belong to the set. In contrast, fuzzy set theory permits the gradual assessment of the membership of elements to a set described by membership function.

Governing Paradigm Shift Availability of data collected through sensing capabilities in urban systems of our time has made revolutionary changes from vertical to horizontal governing.

Graph Databases A graph database is a form of database whereas data is structured as graphs. This makes it possible to query and analyze relationships and to model their attributes. A graph database takes advantage of convenient graph features to manage (i.e., read, delete, update and even create) data and represent it as graph data models. Graph databases are very useful for data that are highly connected, as they can model relationships. They are able to tackle the increasing complexity of data. Thus, graph databases provide a viable alternative to relational databases.

Influence is the capacity to have an effect on the character, development, or behaviour of someone or something, or the effect itself.

Intelligent Urban Governance Intelligent urban governance utilizes data-driven approaches to impact on awareness and behavior of individuals and improves social and institutional collaboration and cooperation in cognitive cities.

Interpretability Interpretability is a property of intelligent systems that enable people to understand the knowledge based embodied in the systems and to comprehend the reasons behind the inferred decisions. Interpretability is multi-faceted as it embodies both structural qualities (also called readability) and semantical aspects (also called comprehensibility). Although not easy to quantify, interpretability is an essential quality for designing intelligent systems for Collaborative Intelligence.

Knowledge Graph A knowledge base containing entities and relationships among them. One of the main applications is to enhance search engines' results with information gathered from a variety of sources and facilitate information retrieval. The first one was popularized by Google, used in its search engine.

Learning Algorithms The aim of a learning algorithm is to adapt a system to a specific input-output transformation task. The algorithm is used to extract patterns from data processing. These patterns in turn enable learning processes in systems, allowing processes to be automated. Learning algorithms can be supervised, unsupervised or reinforcement.

Linguistic Descriptions of Complex Phenomena (LDCP) is aimed at automatic generation of human-like textual descriptions about complex phenomena such as those taking place normally in cities. It follows a human-centric design methodology which has interpretable fuzzy models in the core. LCDP employs fuzzy models and advanced techniques of natural language generation to provide explanations of phenomena that are immediately to understand by people without technical skills.

Linguistic Summary A Linguistic Summary is usually a short quantified sentence of natural language, which express the knowledge in the data in a concise and easily understandable way for variety of data users.

Natural Language Processing (NLP) is a branch of artificial intelligence that helps computers understand, interpret and manipulate human language.

Neo4j is one of the first NoSQL graph databases as it was first introduced in 2003. Due to its Java fundamentals, it offers a versatile basis for creating a graph database.

It is very flexible and useful for tagging, metadata annotations and social networks. Some of its main advantages are that it allows for atomicity, consistency, isolation, durability (ACID) transactions. It is highly available, able to handle billions of nodes and relationships; allows for very fast querying; and has its own graph query language, Cypher. Additionally, Neo4j applies both native graph processing and storage, and it supports PGM as a data model.

Ontological Design Research (ODR) relies on the basic idea that humans shape their world as this world affects and shapes them. ODR is a research model in the making based on ADR, but focusing on the particularities regarding the user, the context and the artefact in complex sociotechnical systems like the cognitive city with lots of subsystems and -contexts, stakeholders, interest groups and individuals. On the one hand, ODR is supposed to emphasize the role of the end-user as co-creators, -designers and -researchers. This is especially important for designing cognitive cities as this aspect touches one of the concept's main promises: the involvement of the citizen in shaping the city. On the other hand, ODR helps to conceptualize the role of artificial intelligence in human-centered systems like the cognitive city: The cognitive city-concept refers to the reciprocity of communication and interaction between city-related ICT and the citizens. Therefore, it is necessary to address the idea that artefacts are not only designed in an iterative process, but that they also have a creative effect in said process.

Ontology An ontology is a specification of a conceptualization. It is a formal naming and definition of the types, properties, and interrelationships of the entities that really exist in a particular domain of discourse, but also the entities and properties that can exist. As an element of the Semantic Web in Artificial Intelligence, an ontology is a model for describing the world that consists of a set of types, properties, and relationship types that should closely resemble the real world.

Ordinary Reasoning Reasoning allows either rejecting that which contradicts some initial information on something or considering what is not in contradiction with it. It is not possible to reason on nothing. From nothing only nothing can be reached. Therefore, reasoning basically consists of conjecturing and refuting from some previous information, evidence, or knowledge.

Personalization In marketing, the term personalization is a means of meeting the customer's needs more effectively and efficiently, making interactions faster and easier and, consequently, increasing customer satisfaction and the likelihood to repeat the visits.

Questionnaire A Questionnaire is a set of questions for obtaining information from respondents. Generally, questions can be ordered-scales, discrete, numerical and open-ended.

Recommendation System A Recommendation System is an information filtering system that seeks to predict the "rating" or "preference" a user would give to an item in order to provide personalized suggestions.

Semantic Ontologies Ontologies are formal vocabularies written by means of Semantic Web standards. Ontologies are stored as documents on the Web. They describe and represent a data domain as a set of concepts and relationships between them. We consider a data domain as a set of related concepts that belong to a specific area of interest. Ontologies are used to represent and relate web resources as a set of related concepts, in order to create a generic knowledge that can be shared across different software applications.

Semantic Web The Semantic Web was defined by Tim Berners-Lee, as "an extension of the current web in which information is given well-defined meaning, better enabling computers and people to work in cooperation". The Semantic Web adds metadata to the information available on the Web, creating vocabularies that describe additional information such as the content, meaning and data relationships. This information should be meaningful and manageable by both humans and computers. To address these aspects, the Semantic Web encompasses a set of standards and technologies used to describe, link, exchange and process data on the Web in a standardized and machine and human-readable way.

Smart City There are many definitions of the term 'smart city'. Most of them have in common the basic idea that the enrichment of city-relevant functions with ICT can contribute to develop efficiently and sustainably the socioecological design of the urban space. The collection and analysis of city-related data as well as the coordination of their use by means of internet and web-based services are intended to help to develop cities into better, more beautiful, more viable places. Smart City-concepts and -projects usually focus on the enhancement of efficiency and sustainability. Hereby, Smart solutions can be applied to lots of city-relevant subjects, for example smart mobility, smart energy, smart environment, smart economy, smart living, and smart governance. By addressing urban complexity and large urban data sets alongside with numerous stakeholders using urban services, cities foster interaction and information sharing across citizens.

Smart Textile is made from fabrics with sensors, i.e., active devices capable of sending or receiving signals to and from the body and, so, acting as real information transducers. The crucial components of smart textiles are sensors and actuators, along with control units to manage their activity. Two types of intelligent textiles are distinguished: passive textiles, capable of detecting the environment, and active textiles, which manage and control different signals from the body. Smart textiles can become a universal and ubiquitous interface, monitoring the wearer and the environment and reacting to external stimuli, whether mechanical, electrical, chemical or thermal. So, they should have a prominent role in cognitive cities.

Social Interaction with Cognitive Maps In cognitive cities, citizens are the focus: how they feel, how they perceive problems, and how they are convinced by experts to modify past behaviors in order to adopt new solutions in a constantly changing, dynamic environment. In this task, some types of social interaction can emerge. Cognitive maps offer a graphical method for explaining causal dependencies between actions or agents, as conflict, cooperation or accommodation. Competition or conflict occurs when two or more individuals or institutions hold opposing views to a problem. Cooperation when they collaborate to better approach an issue and accommodation when there is neither cooperation nor confrontation; i.e., when a person or institution superficially disagrees with another and a consensus solution is possible. Cognitive maps offer a toll for representing social relations and to diagnose causal dependencies between different behaviors.

Social Networks are networks of individuals (such as friends, acquaintances, and coworkers) connected by interpersonal relationships. In technology, it is an online service or site through which people create and maintain interpersonal relationships.

Subjectivity is the quality of being based on or influenced by personal feelings, tastes, or opinions.

T-norm The T-norm is a function for conjunction among imprecise predicates. These functions must comply with certain basic properties: commutative, associative, monotonicity, and border conditions. T-norms have also property of downward reinforcement.

Urban Governance The term urban governance refers to the set of mechanisms for distributing power and allocating resources through the exercising control and coordination processes in a way that fosters partnership among all sectors of the society.